PREVENTING OCCUPATIONAL FALLS

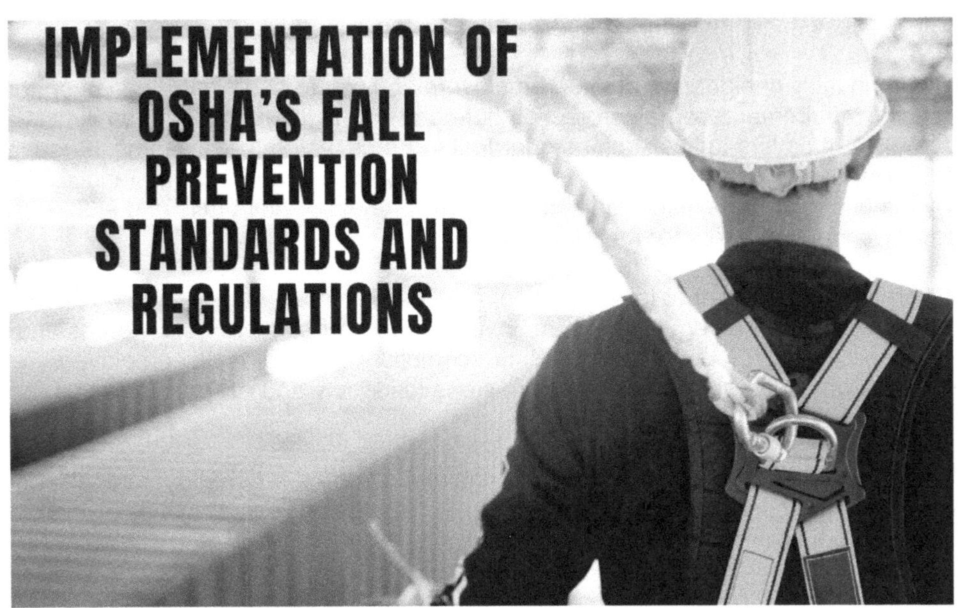

RED DOT PUBLICATIONS

INTRODUCTION

Fall prevention is a critical aspect of workplace safety, particularly in the construction industry where workers are often exposed to heights and potentially hazardous situations. According to the Occupational Safety and Health Administration (OSHA), falls are one of the leading causes of workplace injuries and fatalities. In order to prevent falls and ensure the safety of workers, employers must provide appropriate fall protection measures, including guardrail systems, safety net systems, and personal fall arrest systems.

This manual is designed to provide employers and workers with information on the specific requirements for fall protection during residential construction, as outlined by OSHA. It covers topics such as guardrail systems, safety net systems, personal fall arrest systems, and training resources for fall protection. By understanding and implementing these requirements, employers can help prevent falls and ensure a safer work environment for their employees.

It is important to note that fall protection requirements may vary depending on the jurisdiction and local regulations. Employers should consult OSHA standards and any applicable local regulations to ensure compliance. Additionally, employers should always prioritize safety and take steps to identify and mitigate potential hazards in the workplace.

We hope that this manual will serve as a helpful resource for employers and workers in the construction industry who are committed to ensuring a safe and healthy work environment.

Fall prevention at work

During the construction of residential structures, OSHA requires fall protection to be provided to employees who are working at heights of 6 feet or more above a lower level. The specific requirements include:

1. Guardrail Systems: Guardrail systems are one of the primary methods of fall protection. They consist of a top rail, midrail, and toeboard to prevent workers from falling off the edge. The OSHA requirements for guardrail systems include:

  - The top rail must be 42 inches (plus or minus 3 inches) above the working surface.

  - The midrail should be positioned midway between the top rail and the working surface.

  - The guardrail system should be able to withstand a force of at least 200 pounds applied in any downward or outward direction.

  - The guardrail system must be free from any sharp or protruding objects that could cause injury.

2. Safety Net Systems: Safety net systems are another option for fall protection during residential construction. Safety nets are installed below the working area to catch workers in the event of a fall. OSHA requires safety nets to meet certain specifications, including the ability to catch a falling worker and be installed with enough clearance to prevent contact with the surface or structures below.

3. Personal Fall Arrest Systems: If guardrail systems or safety nets are not feasible or do not provide adequate protection, personal fall arrest systems (PFAS) can be used. PFAS typically consist of an anchorage point, a full-body harness, and a connecting device (such as a lanyard or lifeline) that allows a worker to be safely arrested in the event of a fall.

  - PFAS must be capable of supporting at least 5,000 pounds per employee attached.

  - The maximum allowable free fall distance should not exceed 6 feet.

  - The PFAS must be inspected before each use to ensure proper function and condition.

It is important to note that the specific requirements for fall protection during residential construction may vary depending on the jurisdiction and local regulations. Employers should consult OSHA standards and any applicable local regulations to ensure compliance.

Regarding training resources for fall protection, OSHA provides various guidelines and resources to help employers provide training to their employees. The OSHA Construction Industry Outreach Training Program is a comprehensive training program that covers fall protection and other safety topics. Additionally, OSHA's website offers a wide range of educational materials, including fact sheets, videos, and guides, specifically addressing fall protection training. Employers can also consider hiring qualified trainers or safety consultants who specialize in fall protection to provide on-site training.

Now, let's discuss the difference between guardrail systems and personal fall arrest systems:

1. Guardrail Systems: Guardrail systems are physical barriers erected along the edges of elevated surfaces to prevent falls. They consist of a top rail, midrail, and toeboard. Guardrails provide passive protection, meaning they create a barrier that prevents workers from accessing hazardous areas or falling off edges. They are a collective form of fall protection as they can protect multiple workers simultaneously. Guardrails are typically a permanent or semi-permanent installation and do not require individual workers to wear additional personal protective equipment.

2. Personal Fall Arrest Systems (PFAS): Personal fall arrest systems are designed to arrest or stop a worker's fall should they lose balance or encounter a fall hazard. PFAS typically consist of a full-body harness, a connecting device (such as a lanyard or

lifeline), and an anchorage point. If a worker falls, the connecting device absorbs the energy of the fall and prevents them from hitting the lower level. PFAS provide individual protection and are used when other fall protection measures, such as guardrails, are not feasible or do not provide adequate protection. PFAS require proper training, inspection, and maintenance to ensure their effectiveness.

While guardrail systems provide a physical barrier to prevent falls, personal fall arrest systems are designed to arrest falls and minimize the impact or risk of injuries should a fall occur. The choice between guardrail systems and personal fall arrest systems depends on factors such as the specific work environment, the nature of the tasks being performed, and the level of fall protection required. It is essential to assess the specific hazards and requirements of the work area to determine the appropriate system to use in ensuring worker safety.

During the construction of residential structures, OSHA requires fall protection to be provided to employees who are working at heights of 6 feet or more above a lower level. The specific requirements include:

1. Guardrail Systems: Guardrail systems are one of the primary methods of fall protection. They consist of a top rail, midrail, and toeboard to prevent workers from falling off the edge.

The OSHA requirements for guardrail systems include:

- The top rail must be 42 inches (plus or minus 3 inches) above the working surface.

- The midrail should be positioned midway between the top rail and the working surface.

- The guardrail system should be able to withstand a force of at least 200 pounds applied in any downward or outward direction.

- The guardrail system must be free from any sharp or protruding objects that could cause injury.

2. Safety Net Systems: Safety net systems are another option for fall protection during residential construction. Safety nets are installed below the working area to catch workers in the event of a fall. OSHA requires safety nets to meet certain specifications, including the ability to catch a falling worker and be installed with enough clearance to prevent contact with the surface or structures below.

3. Personal Fall Arrest Systems: If guardrail systems or safety nets are not feasible or do not provide adequate protection, personal fall arrest systems (PFAS) can be used. PFAS typically consist of an anchorage point, a full-body harness, and a connecting device (such as a lanyard or lifeline) that allows a worker to be safely arrested in the event of a fall.

- PFAS must be capable of supporting at least 5,000 pounds per employee attached.

- The maximum allowable free fall distance should not exceed 6 feet.

- The PFAS must be inspected before each use to ensure proper function and condition.

It is important to note that the specific requirements for fall protection during residential construction may vary depending on the jurisdiction and local regulations. Employers should consult OSHA standards and any applicable local regulations to ensure compliance.

Regarding training resources for fall protection, OSHA provides various guidelines and resources to help employers provide training to their employees. The OSHA Construction Industry Outreach Training Program is a comprehensive training program that covers fall protection and other safety topics. Additionally, OSHA's website offers a wide range of educational materials, including fact sheets, videos, and guides, specifically addressing fall protection training. Employers can also consider hiring qualified trainers or safety consultants who specialize in fall protection to provide on-site training.

Now, let's discuss the difference between guardrail systems and personal fall arrest systems:

1. Guardrail Systems: Guardrail systems are physical barriers erected along the edges of elevated surfaces to prevent falls. They consist of a top rail, midrail, and toeboard. Guardrails provide passive protection, meaning they create a barrier that prevents workers from accessing hazardous areas or falling off edges. They are a collective form of fall protection as they can protect multiple workers simultaneously. Guardrails are typically a permanent or semi-permanent installation and do not require individual workers to wear additional personal protective equipment.

2. Personal Fall Arrest Systems (PFAS): Personal fall arrest systems are designed to arrest or stop a worker's fall should they lose balance or encounter a fall hazard. PFAS typically consist of a full-body harness, a connecting device (such as a lanyard or lifeline), and an anchorage point. If a worker falls, the connecting device absorbs the energy of the fall and prevents them from hitting the lower level. PFAS provide individual protection and are used when other fall protection measures, such as guardrails, are not feasible or do not provide adequate protection. PFAS require proper training, inspection, and maintenance to ensure their effectiveness.

While guardrail systems provide a physical barrier to prevent falls, personal fall arrest systems are designed to arrest falls and minimize the impact or risk of injuries should a fall occur. The choice between guardrail systems and personal fall arrest systems depends on factors such as the specific work environment, the nature of the tasks being performed, and the level of fall protection required. It is essential to assess the specific hazards and requirements of the work area to determine the appropriate system to use in ensuring worker safety.

According to OSHA standards, specifically 29 CFR 1926 Subpart M, the specific requirements for fall protection in the construction industry include the following:

1. Fall protection for employees working at heights of 6 feet or more: OSHA requires employers to provide fall protection systems such as guardrail systems, safety net systems, or personal fall arrest systems when employees are working at heights of 6 feet or more above a lower level.

2. Fall protection for employees working on scaffolds: Employees working on scaffolds must be provided with fall protection, which can include guardrail systems, personal fall arrest systems, or a combination of guardrails and personal fall arrest systems.

3. Fall protection for employees working on unprotected edges and sides: Employees working on unprotected sides and edges, such as leading edges, must be protected from falls. This can be achieved through the use of guardrail systems, safety net systems, personal fall arrest systems, or other fall protection measures.

4. Fall protection during construction of residential structures: For employees working on residential structures, fall protection must be provided when working at heights of 6 feet or more. This can include guardrail systems, safety net systems, personal fall arrest systems, or other appropriate measures.

5. Training: Employers are required to provide training to employees who may be exposed to fall hazards. The training should cover the nature of fall hazards, the proper use of fall protection systems, and the limitations of fall protection equipment.

Regarding the sample fall protection plans mentioned in the document, the specific details and contents of the plans are not provided in the given excerpt. However, the document mentions that sample fall protection plans are included for building construction, aviation units, and ground maintenance units. These sample plans serve as examples that can be customized to fit the specific needs and requirements of the respective units or organizations.

As for case studies or examples of successful fall protection programs implemented by organizations, the given excerpt of the document does not provide any specific case studies or examples. The document primarily focuses on providing guidance and information on fall protection programs rather than showcasing specific case studies. To find examples of successful fall protection programs, it may be helpful to research industry-specific case studies, consult industry associations or safety organizations, or look for success stories shared by organizations in the construction, aviation, or ground maintenance sectors.

When selecting between guardrail systems and personal fall arrest systems, employers should consider several key factors:

1. Workplace Hazards: Assess the specific hazards present in the work area. Guardrail systems provide passive protection and are effective in preventing falls in areas with consistent fall hazards. Personal fall arrest systems are more suitable for situations where workers may need to move freely, work at different heights, or encounter intermittent fall hazards.

2. Accessibility and Mobility: Evaluate the need for mobility and accessibility in the work area. Guardrail systems may limit mobility and access to certain areas, while personal fall arrest systems allow workers to move more freely. If workers need to access different levels or areas with limited space, personal fall arrest systems may be more appropriate.

3. Fall Distance and Clearance: Consider the fall distance and clearance requirements. Guardrail systems are effective in preventing falls over edges, while personal fall arrest systems are designed to arrest falls after a certain distance. If the fall distance is significant or there is limited clearance for a worker to fall safely, personal fall arrest systems are typically the preferred choice.

4. Task Variety and Flexibility: Assess the variety of tasks performed in the work area. Guardrail systems are generally suitable for areas with consistent tasks and fall hazards. Personal fall arrest systems offer more flexibility and adaptability for workers who engage in different tasks or move between various work areas.

5. Training and Familiarity: Consider the training requirements and worker familiarity. Guardrail systems are relatively straightforward and require minimal training. Personal fall arrest systems, on the other hand, require proper training on their use, inspection, and maintenance. Employers should ensure that workers using personal fall arrest systems receive the necessary training to understand the equipment, its limitations, and proper procedures.

Examples of situations where personal fall arrest systems would be more appropriate than guardrail systems include:

1. Roofing Work: Roofing work often involves moving along different areas of the roof, where guardrails may not be practical or feasible due to the roof's shape or complexity. Personal fall arrest systems allow workers to move freely while still providing fall protection.

2. Tower Climbing: Workers who climb towers or tall structures need the flexibility to move vertically and horizontally. Personal fall arrest systems provide the necessary fall protection while allowing for greater mobility.

3. Suspended Scaffolding: When working on suspended scaffolding, guardrails may obstruct the movement of workers. Personal fall arrest systems, combined with appropriate anchor points, can provide fall protection while allowing workers to access different areas of the scaffolding.

Regarding training requirements for employees using personal fall arrest systems, OSHA requires employers to provide training that covers the proper use, inspection, and maintenance of the equipment.

The training should include:

- Recognition of fall hazards and the importance of fall protection.

- Understanding the components of a personal fall arrest system.

- Proper donning, adjustment, and use of a full-body harness.

- Selection and inspection of connecting devices (e.g., lanyards, lifelines).

- Anchorage point selection and proper anchoring techniques.

- Procedures for fall arrest system limitations and rescue/recovery.

Employers must ensure that employees who use personal fall arrest systems are trained by a qualified person and demonstrate competency in using the equipment before working at heights or in areas with fall hazards.

When selecting between guardrail systems and personal fall arrest systems, employers should consider several key factors:

1. Nature of Work: Evaluate the specific nature of the work being performed. Consider the tasks, work environment, and potential fall hazards. Determine whether the work involves stationary tasks or requires movement and mobility in different areas.

2. Flexibility and Access: Assess the need for flexibility and access to various parts of the work area. Guardrail systems are fixed barriers that provide continuous fall protection but may restrict access to certain areas. Personal fall arrest systems allow workers to move more freely and access different parts of the work area.

3. Fall Distance and Clearance: Consider the fall distance and clearance requirements. Guardrail systems are designed to prevent falls over edges and provide immediate protection. Personal fall arrest systems are suitable when there is a risk of falls from higher elevations with sufficient clearance for the fall arrest system to activate.

4. Mobility and Dexterity: Evaluate the physical mobility and dexterity required for the work. Guardrail systems do not impede movement, while personal fall arrest systems require workers to wear a harness and connect to an anchorage point, which may slightly restrict mobility and dexterity.

5. Training and Familiarity: Consider the training requirements and worker familiarity. Guardrail systems are generally straightforward and require minimal training. Personal fall arrest systems require proper training on their use, inspection, maintenance, and emergency procedures. Employers should ensure that workers using personal fall arrest systems receive the necessary training to understand the equipment and its limitations.

Examples of situations where personal fall arrest systems would be more appropriate than guardrail systems include:

1. Construction at Heights: In construction activities where workers need to move vertically or horizontally at heights, such as on scaffolding, ladders, or elevated platforms, personal fall arrest systems provide mobility and fall protection while allowing access to different work areas.

2. Maintenance and Repair Work: For tasks involving equipment maintenance, repair work, or inspections, personal fall arrest systems are often preferable as they allow workers to move around machinery or structures while remaining protected from falls.

3. Roofing Work: Roofing tasks often require workers to move across the roof surface, making guardrail systems impractical. Personal fall arrest systems with properly anchored lifelines or anchor points provide the necessary fall protection while allowing workers to move freely.

Specific training requirements for employees using personal fall arrest systems include:

- Recognition of fall hazards and the importance of fall protection.

- Understanding the components of a personal fall arrest system, including harnesses, lanyards, lifelines, and anchor points.

- Proper donning, adjustment, and use of a full-body harness.

- Selection and inspection of connecting devices (e.g., lanyards, retractable lifelines).

- Anchorage point selection and proper anchoring techniques.

- Procedures for fall arrest system limitations, including maximum allowable free fall distance and clearance requirements.

- Emergency procedures, including rescue and self-rescue techniques.

- Regular inspection and maintenance of personal fall arrest equipment.

Employers must ensure that employees who use personal fall arrest systems receive training from a qualified person and demonstrate competency in using the equipment before working at heights or in areas with fall hazards.

When selecting between guardrail systems and personal fall arrest systems, employers should consider the following key factors:

1. Workplace Hazards: Assess the specific hazards present in the work area. Consider the frequency and severity of fall hazards, as well as the layout and design of the workspace. Evaluate whether guardrail systems or personal fall arrest systems would provide more effective protection.

2. Mobility and Access: Determine the level of mobility and access required for the tasks being performed. Guardrail systems may limit movement and access to certain areas, while personal fall arrest systems allow workers to move more freely. Consider whether workers need to reach different levels or areas with limited space.

3. Fall Distance and Clearance: Evaluate the fall distance and clearance requirements. Guardrail systems are effective in preventing falls over edges, while personal fall arrest systems are designed to arrest falls after a certain distance. If the fall distance is significant or there is limited clearance for a worker to fall safely, personal fall arrest systems may be more appropriate.

4. Task Variety and Flexibility: Assess the variety of tasks performed in the work area. Guardrail systems are generally suitable for areas with consistent tasks and fall hazards. Personal fall arrest systems offer more flexibility and adaptability for workers who engage in different tasks or move between various work areas.

5. Training and Familiarity: Consider the training requirements and worker familiarity. Guardrail systems are relatively straightforward and require minimal training. Personal fall arrest systems, on the other hand, require proper training on their use, inspection, and maintenance. Employers should ensure that workers using personal fall arrest systems receive the necessary training to understand the equipment, its limitations, and proper procedures.

Examples of situations where personal fall arrest systems would be more appropriate than guardrail systems include:

1. Elevated Platforms: In situations where workers are performing tasks on elevated platforms or work surfaces without suitable guardrails, personal fall arrest systems can be used to provide fall protection while allowing workers to move freely.

2. Construction Work: During construction activities, workers may need to access different levels, work on scaffolding, or move along structural elements. Personal fall arrest systems provide the necessary fall protection and mobility in these dynamic work environments.

3. Aerial Work: When workers are performing tasks from lifts, aerial platforms, or suspended equipment, personal fall arrest systems can provide continuous fall protection as workers move and perform their tasks.

Regarding training requirements for employees using personal fall arrest systems, employers must provide training that covers the proper use, inspection, and maintenance of the equipment. The training should include:

- Recognition of fall hazards and the importance of fall protection.

- Understanding the components of the personal fall arrest system, including harnesses, lanyards, lifelines, and anchor points.

- Proper donning, adjustment, and use of a full-body harness.

- Selection and inspection of connecting devices (e.g., lanyards, lifelines).

- Anchorage point selection and proper anchoring techniques.

- Procedures for fall arrest system limitations and rescue/recovery.

Employers must ensure that employees who use personal fall arrest systems are trained by a qualified person and demonstrate competency in using the equipment before working at heights or in areas with fall hazards.

When selecting between guardrail systems and personal fall arrest systems, employers should consider the following key considerations:

1. Fall Hazard Assessment: Conduct a thorough assessment of the workplace to identify potential fall hazards. Evaluate the height of the work area, the nature of the tasks performed, and the frequency of access required. This assessment will help determine the most suitable fall protection system.

2. Passive vs. Active Protection: Guardrail systems provide passive protection by creating a physical barrier, while personal fall arrest systems require active engagement from the worker. Consider the level of worker involvement and the reliability of each system in preventing falls.

3. Accessibility and Mobility: Assess the need for access and mobility in the work area. Guardrail systems may restrict access to certain areas, while personal fall arrest systems allow workers to move more freely. Consider the tasks that require workers to move horizontally or vertically.

4. Fall Clearance and Distance: Evaluate the fall clearance and distance requirements. Guardrail systems are effective in preventing falls over edges, while personal fall arrest systems are designed to arrest falls after a certain distance. If there is limited clearance or the fall distance is significant, personal fall arrest systems may be more appropriate.

5. Training and Familiarity: Consider the training requirements and worker familiarity with the systems. Guardrail systems are relatively straightforward and require minimal training. Personal fall arrest systems, on the other hand, necessitate training on proper use, inspection, maintenance, and emergency procedures. Ensure that workers using personal fall arrest systems receive appropriate training.

Examples of situations where personal fall arrest systems would be more appropriate than guardrail systems include:

1. Elevated Work on Structures: When workers are performing tasks on elevated structures, such as towers, bridges, or communication poles, personal fall arrest systems allow them to move freely while providing continuous fall protection.

2. Scaffolding and Temporary Platforms: In situations where workers are using scaffolding or temporary platforms, personal fall arrest systems are more suitable as they allow workers to move across different levels and access various areas.

3. Maintenance and Repair Work: For tasks involving maintenance or repair work on machinery, equipment, or infrastructure, personal fall arrest systems provide flexibility and fall protection while allowing workers to navigate around obstacles.

Training requirements for employees using personal fall arrest systems include:

- Recognition of fall hazards and the importance of fall protection.

- Understanding the components of a personal fall arrest system, including harnesses, lanyards, anchor points, and connectors.

- Proper donning, adjustment, and use of a full-body harness.

- Selection, inspection, and maintenance of connecting devices and anchor points.

- Fall arrest system limitations, including maximum allowable free fall distance and clearance requirements.

- Emergency procedures, including rescue and self-rescue techniques.

- Hands-on training and practice in using the equipment.

Employers must ensure that employees receive training from a qualified person and demonstrate competency in using the personal fall arrest systems before performing work at heights or in areas with fall hazards. Regular refresher training may also be necessary to maintain proficiency.

The purpose of this guide is to assist unit leaders, Army safety and occupational health professionals, Soldiers and Army Civilians in complying with the requirements of OSHA's Fall Protection Standard, 29 CFR 1926 (Construction), Subpart M and 29 CFR 1910 (General Industry), Subpart D, as well as to provide other helpful information to manage risk and sustain readiness. It is not intended to supersede the requirements of the standard. Fall protection is also covered under 29 CFR 1915 (Shipyard), 29 CFR 1917 (Marine Terminals) and 29 CFR 1918 (Longshoring). Users of this document should review the standard for particular requirements that are applicable to their individual situations and make adjustments to this program that are specific to their units. It is highly recommended they add information relevant to their particular facility or organization to develop an effective and comprehensive program.

29 CFR 1926 and 29 CFR 1910
Fall Prevention Program

Fall Prevention Program Overview

The following Fall Prevention Program is provided only as a guide to assist units and Soldiers/employees in complying with the requirements of OSHA's Fall Protection Standard, 29 CFR 1926 (Construction), Subpart M and 29 CFR 1910 (General Industry), Subpart D, as well as to provide other helpful information. It is not intended to supersede the requirements of the standard. Fall protection is also covered under 29 CFR 1915 (Shipyard), 29 CFR 1917 (Marine Terminals) and 29 CFR 1918 (Longshoring). Units should review the standard for particular requirements that are applicable to their individual situation and make adjustments to this program that are specific to their unit. Units will need to add information relevant to their particular facility in order to develop an effective, comprehensive program.

Instructions for Written Fall Protection Plan Development

In each of the applicable sections below, fill in the items required in "red" text in the underlined areas. They will aid in developing a site-specific Fall Protection Plan for your organization. Customized procedures may be added as appropriate. The "Responsible Person" may not be the same individual in each instance, select the individual or office according to the requirement. Any specific topics not included in the plan can be added as an additional appendix.

Upon completion of the applicable template, all "red" text should be changed to "black" text due to review and acknowledgment that the method, technique or principle applies to your unit's operations and you have customized the text to align with site-specific implementation. It also means you have to exercise judgment to implement an appropriate and effective program.

The templates have been prepared using information and data from sources considered technically reliable and is believed to be correct. The USACRC makes no warranties, expressed or implied, as to the accuracy of the information contained herein. The USACRC cannot anticipate all conditions under which this information and the subject products may be used and the actual conditions of use are beyond its control. The user is responsible for evaluating all available information when using the subject product for any particular use and complying with all federal, state, and local laws, directives, statutes and regulations. The use of trademarked name(s) does not imply endorsement by the U.S. Army but is intended only to assist in the identification of a specific product.

Table of Contents

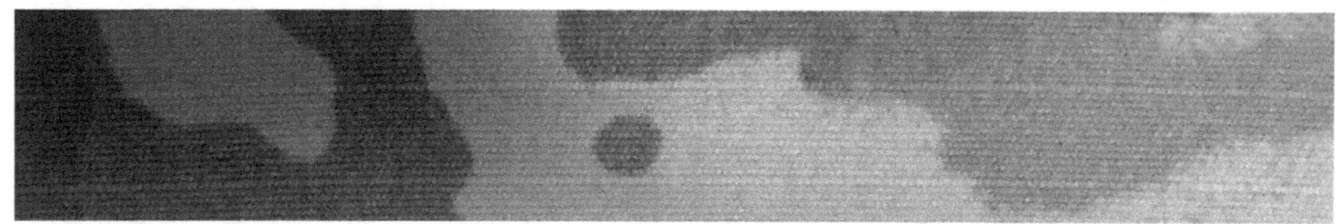

Chapter 1 –
FALL PREVENTION PROGRAM OVERVIEW

Purpose. This Guide establishes criteria and requirements for developing and managing Fall Prevention programs to protect all Army personnel (military and civilians) at Army activities.

Falls are the leading cause of work-related injuries and fatalities in construction accidents nationwide and are ranked second in general industry. According to the Bureau of Labor Statistics, most work-related injuries and fatalities are on the decline. However, the number of fall-related injuries and fatalities is increasing, accounting for more than 13% of the total number of fatal work injuries. In the United States, approximately three fall-related fatalities occur each working day.

From an Army perspective, injuries to Soldiers and civilians sustained from falls can significantly impact resources and hinder mission capability. Protecting the workforce is a responsibility shared by everyone, at all levels of the organization. However, it is you, the leader, who makes a unique contribution to job safety in that you are aware of the skills, physical condition, capabilities, and limitations of your people. You know the job, and have the authority to inspect, correct, and direct. No one is in a better position to prevent accidental falls in the workplace than you. This Leader's Guide to Fall Protection safety guide is designed to establish criteria for fall protection programs in order to heighten awareness and protect Army personnel exposed to fall hazards in the workplace. To simplify the development of a fall protection program, portions of this guide are highlighted in red. Simply insert the required information to develop a basic written unit Fall Prevention Program.

Objective. The objective of the Army Fall Prevention Program is to identify and evaluate fall hazards to which Soldiers/employees will be exposed, and to provide specific training as required by the Occupational Safety and Health Administration (OSHA) Fall Protection Standard, 29 CFR 1926, Subpart M and Walking Working Surfaces Standard, 29 CFR 1910, Subpart D.

Policy. It is the policy of the Army to protect its Soldiers/employees from occupational injuries by implementing and enforcing safe work practices and appointing a Competent Person(s) to manage the Fall Protection Program (See Appendix G). The Army Fall Protection Program shall comply with the OSHA requirements. A copy of the OSHA Fall Protection Standard shall be made available to all Soldiers/employees, and may be obtained from the designated Responsible Person. In accordance with Department of the Army Pamphlet (DA Pam) 385-10, Army Safety Program, every facility having personnel working at heights, exposed to fall hazards and using fall protection equipment is responsible for establishing, implementing and managing a Fall Protection Program, which includes identification and elimination or control of fall hazards. These facilities are responsible for assigning responsibilities; surveying and assessing fall hazards; providing prevention and control measures; training of personnel; inspecting fall protection equipment; auditing and evaluating the effectiveness of the program; and ensuring proper installation and use of fall protection systems and providing rescue equipment with accompanying rescue procedures. Fall protection must be provided to personnel exposed to fall hazards on any elevated walking/working surface with unprotected side, edge, or floor openings, from which there is a possibility of falling 4 feet or more (General Industry Standard 29 CFR 1910) or 6 feet or more (Construction Standard 29 CFR 1926) to a lower level or where there is a possibility of a fall from any height onto dangerous equipment, into a hazardous environment, or onto an impalement hazard. A written Fall Protection Plan is required for each worksite exposing Soldiers/employees to fall hazards. Information on written plans for building construction is available in Appendix A and B; Aviation maintenance in Appendix H, I, J and K; and Ground maintenance in Appendix L, M, and N.

Chapter 2 –
ASSIGNMENT OF RESPONSIBILITY

Employer: It is the responsibility of the organization to provide fall protection to affected Soldiers/employees and to ensure all Soldiers/employees understand and adhere to the procedures of this plan and follow the instructions of the designated Responsible Person.

Program Manager: It is the responsibility of the designated Responsible Person as the FPPM to implement this program by:

- Performing routine safety checks of work operations. Examples of Fall Hazard Survey Reports are available in Appendix B, J and N.
- Enforcing the organization's safety policy and procedures.
- Correcting any unsafe practices or conditions immediately.
- Training Soldiers/employees and supervisors in recognizing fall hazards and the use of fall protection systems.
- Maintaining records of Soldiers/employees training, equipment issue, and fall protection systems used at the organization job sites.
- Investigating and documenting all incidents that result in Soldiers/employees injury.

Soldiers/employees: It is the responsibility of all Soldiers/employees to:

- Understand and adhere to the procedures outlined in this Fall Protection Program;
- Follow the instructions of the designated Responsible Person;
- Bring to management's attention any unsafe or hazardous conditions or practices that may cause injury to either themselves or any other Soldiers/employees; and
- Report any incident that causes injury to Soldiers/employees, regardless of the nature of the injury.

Statistics show the majority (67%) of falls happen on the same level resulting from slips and trips. The remaining 30% are falls from a height.

Chapter 3 – FALL PROTECTION

Slips, Trips and Falls. Identifying fall hazards and deciding how best to protect workers is the first step in reducing or eliminating fall hazards. People have fallen from considerable heights and received only a few broken bones, while others fall to the floor from a standing or sitting position and die from their injuries. Nearly all falls result from conditions or practices that seem obvious; however, preventing such accidents requires maintaining safe conditions in the workplace and training to ensure safe actions by employees.

Housekeeping. Good housekeeping is the first and the most important (fundamental) level of preventing falls due to slips and trips. It includes:

- Cleaning all spills immediately
- Marking spills and wet areas
- Mopping or sweeping debris from floors
- Removing obstacles from walkways and always keeping walkways free of clutter
- Securing (tacking, taping, etc.) mats, rugs and carpets that do not lay flat
- Always closing file cabinet or storage drawers
- Covering cables that cross walkways
- Keeping working areas and walkways well lit
- Replacing used light bulbs and faulty switches

Without good housekeeping practices, any other preventive measures such as installation of sophisticated flooring, specialty footwear or training on techniques of walking and safe falling will never be fully effective.

Tripping and Stumbling Hazards. Trips happen when your foot collides (strikes, hits) with an object causing you to lose balance and eventually fall. Common causes of tripping are:

- Objects or materials in walkways
- Projecting parts of machines or equipment
- Equipment or material on stairs
- Scrap or waste materials
- Pipe or conduit set near floor level
- Extension cords, power cables, air hoses, welding cables, fuel, gas and oxygen hoses
- Obstructed view
- Poor lighting
- Wrinkled carpeting
- Bottom drawers not being closed
- Uneven (steps, thresholds) walking surfaces

Slipping Hazards. Slips happen where there is too little friction or traction between the footwear and the walking surface. Common causes of slips are:

- Wet or oily surfaces
- Occasional spills
- Weather hazards
- Loose, unanchored rugs or mats
- Flooring or other walking surfaces that do not have some degree of traction in all areas

Unstable Surfaces. The movement of vehicles during maintenance operations present a serious fall hazard.

- Chock blocks will be used whenever a vehicle or trailer is parked or maintenance is being conducted.

- Ensure chock blocks are placed behind the
  tire in the direction of downward travel.
- Remind Soldiers/employees to make sure brakes
  are set, the engine is off, and at least one wheel
  is choked during loading and unloading.
- Ensure parking brakes are serviceable; report any brake
  malfunctions to maintenance personnel immediately.

Falls from the Same Level

Working Surface. Unnoticed changes in surface friction
are implicated in many accidents. Going from a less
slippery floor to a more slippery one produces slips;
the opposite change produces trips and missteps.
These unnoticed changes can be reduced by:

- Ensuring that different surface materials or
  coatings have transition zones between them.
- Clearly marking any surface where friction changes.
- Using good housekeeping procedures to reduce
  changes in surface friction caused by spills,
  worn spots, and loose or irregular floors.

The above recommendations are of particular importance
in manual materials handling where any handling
other than direct lifting involves horizontal inertial
forces transmitted from the container to the body. Such
forces require increased frictional forces to prevent
foot slippage. Carrying weights also affects the body's
learned reflexes for recovering from a slip or trip. In
such situations, the body's normal weight distribution
is altered and the arms are prevented from being used
to regain balance or recover from another moving mass
in close proximity to the falling operator. There is the
potential for both crushing and puncturing the body.

Stairways and Ladders. Stairways and ladders are a major
source of injuries and fatalities among workers. OSHA
estimates that there are 24,882 injuries and as many
as 36 fatalities per year due to falls from stairways and
ladders used in construction. Nearly half of these injuries
are serious enough to require time off the job—11,570
lost workday injuries and 13,312 non-lost workday
injuries occur annually due to falls from stairways and
ladders used in construction. This data demonstrates
that work on and around ladders and stairways is

hazardous. More importantly, it shows that compliance
with OSHA requirements for the safe use of ladders and
stairways could have prevented many of these injuries.

The OSHA rules apply to all stairways and ladders
used in industry, alteration, repair (including painting
and decorating), and demolition of work sites covered
by OSHA safety and health standards. They also
specify when stairways and ladders must be provided.
They do not apply to ladders that are specifically
manufactured for scaffold access and egress,
but they do apply to job-made and manufactured
portable ladders intended for general purpose use
that are then used for scaffold access and egress.

General Requirements. The following general requirements
apply to construction covered under 29 CFR Part 1926:

- A stairway or ladder must be provided at all worker
  points of access where there is a break in elevation
  of 19 inches (48 cm) or more and no ramp, runway,
  sloped embankment or personnel host is provided.
- When there is only one point of access between
  levels, it must be kept clear to permit free passage
  by workers. If free passage becomes restricted, a
  second point of access must be provided and used.
- All stairway and ladder fall protection systems
  required by these rules must be installed and all
  duties required by the stairway and ladder rules
  must be performed before Soldiers/employees begin
  work that requires them to use stairways or ladders
  and their respective fall protection systems.

Stairways. The following general requirements apply to
all stairways used in construction and general industry:

- Stairways that will not be a permanent part
  of the structure on which construction work
  is performed must have landings at least 30
  inches deep and 22 inches wide (76 x 56 cm) at
  every 12 feet (3.7 m) or less of vertical rise.
- Stairways must be installed at least 30 degrees, and
  no more than 50 degrees from the horizontal.
- Variations in riser height or stair tread depth
  must not exceed 1/4 inch in any stairway
  system, including any foundation structure
  used as one or more treads of the stairs.

- Where door or gates open directly onto a stairway, a platform must be provided that is at least 20 inches (51 cm) in width beyond the swing of the door.
- Metal pan landings and metal pan treads must be secured in place before filling.
- All stairway parts must be free of dangerous projections such as protruding nails.
- Slippery conditions on stairways must be corrected.
- Spiral stairways that will not be a permanent part of the structure may not be used by workers.

Stairs in Temporary Service. The following requirements apply to stairs in temporary service during construction:

- Except during construction of the actual stairway, stairways with metal pan landings and treads must not be used where the treads and/or landings have not been filled in with concrete or other material, unless the pans of the stairs and/or landings are temporarily filled in with wood or other material. All temporary treads and landings must be replaced when worn below the top edge of the pan.
- Except during construction of the actual stairway, skeleton metal stairs must not be used (where treads and/or landings are to be installed at a later date) unless the stairs are fitted with secured temporary treads and landings.
- Temporary treads must be made of wood or other solid material and installed the full width and depth of the stair.

Stair Rails and Handrails. The following general requirements apply to all stair rails and handrails:

- Every flight of stairs having four or more risers must be equipped with standard stair railings or standard handrails. See 1910.23(d)(1).
- Winding or spiral stairways must be equipped with a handrail to prevent using areas where the tread width is less than 6 inches (15 cm).
- Stair rails installed after March 15, 1991, must not be less than 36 inches (91.5 cm) in height.
- Screens, mesh, intermediate vertical members or equivalent intermediate structural members must be provided between the top rail and stairway steps of the stair rail system.
- Screens or mesh, when used, must extend from the top rail to the stairway step and along the opening between top rail supports.
- Mid-rails, when used, must be located midway between the top of the stair rail system and the stairway steps.

- Intermediate vertical members, such as balusters, when used, must not be more than 19 inches (48 cm) apart.
- Other intermediate structural members, when used, must be installed so that there are no openings of more than 19 inches (48 cm) wide.
- Handrails and the top rails of the stair rail systems must be capable of withstanding, without failure, at least 200 pounds (890 n) of weight applied within 2 inches (5 cm) of the top edge in any downward or outward direction, at any point along the top edge.
- The height of handrails must not be more than 37 inches (94 cm) or less than 30 inches (76 cm) from the upper surface of the handrail to the surface of the tread.
- Stair rail systems and handrails must be surfaced to prevent injuries such as punctures or lacerations and to keep clothing from snagging.
- Handrails must provide an adequate handhold for Soldiers/employees to grasp to prevent falls.
- The ends of stair rail systems and handrails must be constructed to prevent dangerous projections such as rails protruding beyond the end posts of the system.
- Temporary handrails must have a minimum clearance of 3 inches (8 cm) between the handrail and walls, stair rail systems and other objects.
- Unprotected sides and edges of stairway landings must be provided with standard 42-inch (1.1 m) guardrail systems.

Ladders. The following general requirements apply to all ladders, including job-made ladders:

- A double-cleated ladder or two or more ladders must be provided when ladders are the only way to enter or exit a work area having 25 or more Soldiers/employees or when a ladder serves simultaneous two-way traffic. See 1926.1051(a)(2).
- Ladder rungs, cleats and steps must be parallel, level and uniformly spaced when the ladder is in position for use.
- Trestle ladders or extension sections or base sections of extension trestle ladders must not be more than 20 feet in length. See 1910.25(c)(3)(v).
- The spacing of rungs and steps must be on 12-inch centers. Rungs and steps must be corrugated, knurled, dimpled, coated with skid-resistant material or otherwise treated to minimize the possibility of slipping. See 1910.26(a)(1)(v) and 1910.27(b)(1)(ii).
- Ladders must not be tied or fastened together to create longer sections unless they are specifically designed for such use.

- A metal spreader or locking device must be provided on each stepladder to hold the front and back sections in an open position when the ladder is being used.
- Two or more separate ladders used to reach an elevated work area must be offset with a platform or landing between the ladders, except when portable ladders are used to gain access to fixed ladders.
- Ladder components must be surfaced to prevent injury from punctures or lacerations and to prevent snagging of clothing.
- Wood ladders must not be coated with any opaque covering, except for identification or warning labels, which may be placed only on one face of a side rail.

Portable Ladders.

- Non-self-supporting and self-supporting portable ladders must support at least four times the maximum intended load; extra heavy-duty type 1A metal or plastic ladders must sustain 3.3 times the maximum intended load. The ability of a self-supporting ladder to sustain loads must be determined by applying the load to the ladder in a downward vertical direction. The ability of a non-self-supporting ladder to sustain loads must be determined by applying the load in a downward vertical direction when the ladder is placed at a horizontal angle of 75.5 degrees. See 1926.1053(a)(1)(ii).
- The user should equip all portable rung ladders with nonslip bases when there is a hazard of slipping. Nonslip bases are not intended as a substitute for care in safely placing, lashing, or holding a ladder that is being used upon oily, metal, concrete or slippery surfaces. See 1910.25(d)(2)(xix).
- The minimum width between side rails for portable metal ladders must be 12 inches.
- The rungs and steps of portable metal ladders must be corrugated, knurled, dimpled, coated with skid-resistant material or treated to minimize slipping. See 1910.26(a)(1)(v).

Fixed Ladders.

- The minimum design live load must be a single concentrated load of 200 pounds. The number and position of additional concentrated live-load units of 200 pounds each as determined from anticipated usage of the ladder must be considered in the design. See 1910.27(a)(1)(i) and 1910.27(a)(1)(ii).
- The live loads imposed by persons occupying the ladder must be considered to be concentrated at such points as will cause the maximum stress in the structural member being considered. See 1910.27(a)(1)(iii).
- The side rails of through or side-step ladder extensions must extend 42 inches (1.1 m) above parapets and landings. For through ladder extensions, the rungs must be omitted from the extension and must not have less than 18 or more than 24 inches clearance between rails. For side-step or offset fixed ladder sections, at landings, the side rails and rungs must be carried to the next regular rung beyond or above the 42-inch minimum. See 1910.27(d)(3).
- Design stresses for wood components of ladders must not exceed those specified in 1910.25. All wood parts must be free from sharp edges and splinters; sound and free from accepted visual inspection from shake, wane compression failures, decay or other irregularities. Low density wood must not be used. See 1910.25(b)(1)(i) and 1910.27(a)(2).
- The minimum clear length of rungs or cleats must be 16 inches (41 cm). The distance between rungs, cleats and steps must not exceed 12 inches and must be uniform throughout the length of the ladder. See 1910.27(b)(1)(iii) and 1910.27(b)(1)(ii).
- The rungs and steps of fixed metal ladders manufactured after March 15, 1991, must be corrugated, knurled, dimpled, coated with skid-resistant material or treated to minimize slipping. See 1926.1053(a)(6)(i).
- The minimum perpendicular clearance between the centerline of fixed ladder rungs, cleats, steps and any obstruction on the climbing side of the ladder must be 30 inches (76 cm). When unavoidable obstructions are encountered, the distance may be reduced to 24 inches (61 cm), provided that a deflection device is installed to guide Soldiers/employees around the obstruction.
- The step-across distance from the nearest edge of ladder to the nearest edge of equipment or structure must not be more than 12 inches, or less than 21/2 inches. See 1910.27(c)(6).
- A clear width of at least 15 inches (38 cm) must be provided each way from the centerline

of the ladder in the climbing space, except
when cages or wells are necessary.

- Where the total length of a climb on a fixed ladder
  equals or exceeds 24 feet (7.3 m), one of the following
  requirements must be met: fixed ladders must be
  equipped with either (a) ladder safety devices; (b) self-
  retracting lifelines with rest platforms at intervals not
  to exceed 150 feet (45.7 m); or (c) a cage or well, and
  multiple ladder sections, each ladder section not to
  exceed 50 feet (15.2 m) in length. Ladder sections must
  be offset from adjacent sections, and landing platforms
  must be provided at maximum intervals of 50 feet (15.2
  m). (This applies to construction. See 1926.1053(a)(19).
  For general industry requirements see 1910.27(d)(2).)
- The side rails of through or side-step fixed ladders must
  extend 42 inches (1.1 m) above the top level or landing
  platform served by the ladder. For a parapet ladder,
  the access level must be at the roof if the parapet
  is cut to permit passage through it. If the parapet is
  continuous, the access level is the top of the parapet.

Cages for Fixed Ladders.

- Horizontal bands must be fastened to the side rails
  of rail ladders or directly to the structure, building
  or equipment for individual-rung ladders.
- Vertical bars must be on the inside of the horizontal
  bands and must be fastened to them.
- Cages must not extend less than 27 inches
  (68 cm) or more than 30 inches (76 cm) from
  the centerline of the step or rung and must
  not be less than 27 inches (68 cm) wide.
- The inside of the cage must be clear of projections.
- Horizontal bands must be spaced at
  intervals not more than 4 feet (1.2 m) apart
  measured from centerline to centerline.
- Vertical bars must be spaced at intervals
  not more than 9.5 inches (24 cm) apart
  measured from centerline to centerline.
- The bottom of the cage must be between 7 feet (2.1
  m) and 8 feet (2.4 m) above the point of access to the
  bottom of the ladder. The bottom of the cage must
  be flared not less than 4 inches (10 cm) between the
  bottom horizontal band and the next higher band.
- The top of the cage must be a minimum of 42 inches
  (1.1 m) above the top of the platform or the point of
  access at the top of the ladder. Provisions must be made
  for access to the platform or other point of access.

Wells for Fixed Ladders.

- Wells must completely encircle the ladder.
- Wells must be free of projections.
- The inside face of the well on the climbing side of the
  ladder must extend between 27 inches (68 cm) and 30
  inches (76 cm) from the centerline of the step or rung.
- The inside width of the well must be
  at least 30 inches (76 cm).
- The bottom of the well above the point of
  access to the bottom of the ladder must be
  between 7 feet (2.1 m) and 8 feet (2.4 m).

Ladder Safety Devices and Related Support Systems for Fixed Ladders.

- All safety devices must be capable of withstanding,
  without failure, a drop test consisting of a 500-pound
  weight (226 kg) dropping 18 inches (41 cm).
- All safety devices must permit the worker to
  ascend or descend without continually having
  to hold, push or pull any part of the device,
  leaving both hands free for climbing.
- All safety devices must be activated within 2
  feet (.61 m) after a fall occurs and must limit
  the descending velocity of Soldiers/employees
  to 7 feet/second (2.1 m/sec) or less.
- The connection between the carrier or lifeline and
  the point of attachment to the body belt or harness
  must not exceed 9 inches (23 cm) in length.

Mounting Ladder Safety Devices for Fixed Ladders.

- Mountings for rigid carriers must be attached at each
  end of the carrier, with intermediate mountings,
  spaced along the entire length of the carrier, to provide
  the necessary strength to stop workers' falls.
- Mountings for flexible carriers must be attached at
  each end of the carrier. Cable guides for flexible carriers
  must be installed with a spacing between 25 feet (7.6
  m) and 40 feet (12.2 m) along the entire length of
  the carrier, to prevent wind damage to the system.
- The design and installation of mountings and cable
  guides must not reduce the strength of the ladder.
- Side rails and steps or rungs for side-step fixed
  ladders must be continuous in extension.

Use of All Ladders.

- When portable ladders are used for access to an upper
  landing surface, the side rails must extend at least 3
  feet (.9 m) above the upper landing surface. When such

an extension is not possible, the ladder must be secured, and a grasping device such as a grab rail must be provided to assist workers in mounting and dismounting the ladder. A ladder extension must not deflect under a load that would cause the ladder to slip off its support.
- Ladders must be maintained free of oil, grease and other slipping hazards.
- Ladders must not be loaded beyond the maximum intended load for which they were built or beyond their manufacturer's rated capacity.
- Ladders must be used only for the purpose for which they were designed.
- Non-self-supporting ladders must be used at an angle where the horizontal distance from the top support to the foot of the ladder is approximately one-quarter of the working length of the ladder. Wood job-made ladders with spliced side rails must be used at an angle where the horizontal distance is one-eighth the working length of the ladder.
- Fixed ladders must be used at a pitch no greater than 90 degrees from the horizontal, measured from the back side of the ladder.
- Ladders must be used only on stable and level surfaces unless secured to prevent accidental movement.
- Ladders must not be used on slippery surfaces unless secured or provided with slip-resistant feet to prevent accidental movement. Slip-resistant feet must not be used as a substitute for the care in placing, lashing or holding a ladder upon slippery surfaces.
- The area around the top and bottom of the ladders must be kept clear.
- The top of a non-self-supporting ladder must be placed with two rails supported equally unless it is equipped with a single support attachment.
- Ladders must not be moved, shifted or extended while in use.
- Ladders must have nonconductive side-rails if they are used where the worker or the ladder could contact exposed energized electrical equipment.
- Cross-bracing on the rear section of stepladders must not be used for climbing unless the ladders are designed and provided with steps for climbing on both front and rear sections.
- Ladders must be inspected by a Competent Person for visible defects on a periodic basis and after any incident that could affect their safe use.
- Single-rail ladders must not be used.
- When ascending or descending a ladder, the worker must face the ladder.
- Each worker must use at least one hand to grasp the ladder when moving up or down the ladder.
- A worker on a ladder must not carry any object or load that could cause the worker to lose balance and fall.

Structural Defects.
- Portable ladders with structural defects, such as broken or missing rungs, cleats, or steps, broken or split rails, corroded components, or other faulty or defective components, must immediately be marked defective or tagged with "Do not Use" or similar language and withdrawn from service until repaired.
- Fixed ladders with structural defects—such as broken or missing rungs, cleats, or steps, broken or split rails, or corroded components—must be withdrawn from service until repaired.
- Defective fixed ladders are considered withdrawn from use when they are (a) immediately tagged with "Do not Use" or similar language; (b) marked in a manner that identifies them as defective; or (c) blocked (such as with a plywood attachment that spans several rungs).
- Ladder repairs must restore the ladder to a condition meeting its original design criteria before the ladder is returned to use.

Training Requirements. Under the provisions of the standard, employers must provide a training program for each Soldier/employee using ladders and stairways. The program must enable each Soldier/employee to recognize hazards related to ladders and stairways and to use proper procedures to minimize these hazards. For example, employers must ensure that each Soldier/employee is trained by a Competent Person in the following areas, as applicable:
- The nature of fall hazards in the work area
- The correct procedures for erecting, maintaining and disassembling the fall protection systems to be used
- The proper construction, use, placement and care in handling of all stairways and ladders

- The maximum intended load-carrying
  capacities of ladders used

In addition, retraining must be provided for each
Soldier/employee, as necessary, so that the Soldier/
employee maintains the understanding and knowledge
acquired through compliance with the standard.

Chapter 4 – FALL PROTECTION SYSTEMS

Covers.
- All covers shall be secured to prevent
  accidental displacement.
- Covers shall be color-coded or bear the
  markings "HOLE" or "COVER."
- Covers located in roadways shall be able to
  support twice the axle load of the largest
  vehicle that might cross them.
- Covers shall be able to support twice the
  weight of Soldiers/employees, equipment,
  and materials that might cross them.

Guardrail Systems. Guardrail systems shall be
erected at unprotected edges, ramps, runways,
or holes where it is determined by the designated
Responsible Person that erecting such systems will
not cause an increased hazard to Soldiers/employees.
The following specifications will be followed in the
erection of guardrail systems. Top rails shall be:
- At least ¼ inch in diameter (steel or plastic banding is
  unacceptable).
- Flagged every 6 feet or less with a high visibility material
  if wire rope is used.
- Inspected by the designated Responsible Person as
  frequently as necessary to ensure strength and stability.
- 42 inches (plus or minus 3 inches) above the walking/
  working level.
- Adjusted to accommodate the height of stilts, if they are
  in use.
- Midrails, screens, mesh, intermediate vertical members,
  and solid panels shall be erected in accordance with the
  OSHA Fall Protection Standard.
- Gates or removable guardrail sections shall be placed
  across openings of hoisting areas or holes when they are
  not in use to prevent access.

Personal Fall Arrest Systems. Personal fall arrest
systems shall be issued to and used by Soldiers/
employees as determined by the designated Responsible
Person and may consist of anchorage, connectors,
body harness, deceleration device, lifeline, or suitable
combinations. See Appendix N for information on Tie-
Off Considerations and Selection of Safe Anchorages.

- A Fall-arrest System and Equipment Checklist is located
  in Appendix D. Personal fall arrest systems shall:
  - Limit the maximum arresting force to 1,800 pounds;
  - Be rigged so a Soldier/employee cannot free fall
    more than 6 feet or contact any lower level;
  - Bring a Soldier/employee to a complete
    stop and limit the maximum deceleration
    distance traveled to 3½ feet;
  - Be strong enough to withstand twice the
    potential impact energy of a Soldier/employee
    free falling 6 feet (or the free fall distance
    permitted by the system, whichever is less);
  - Be inspected prior to each use for
    damage and deterioration; and
  - Be removed from service if any damaged
    components are detected.

- All components of a fall arrest system shall meet
  the specifications of the OSHA Fall Protection
  Standard, and shall be used in accordance
  with the manufacturer's instructions.
  - The use of non-locking snap-hooks is prohibited.
  - D-rings and locking snap-hooks shall:
    - Have a minimum tensile strength of 5,000 pounds.
    - Be proof-tested to a minimum tensile load
      of 3,600 pounds without cracking, breaking,
      or suffering permanent deformation.

  - Lifelines shall be:
    - Designed, installed, and used under the supervision
      of the designated Responsible Person.
    - Protected against cuts and abrasions.
    - Equipped with horizontal lifeline connection
      devices capable of locking in both directions on
      the lifeline when used on suspended scaffolds
      or similar work platforms that have horizontal
      lifelines that may become vertical lifelines.

  - Self-retracting lifelines and lanyards
    must have ropes and straps (webbing)
    made of synthetic fibers, and shall:
    - Sustain a minimum tensile load of 3,600 pounds if
      they automatically limit free fall distance to 2 feet.

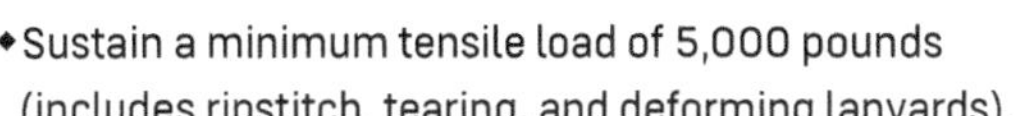

◆ Sustain a minimum tensile load of 5,000 pounds (includes ripstitch, tearing, and deforming lanyards).

- Anchorages must support at least 5,000 pounds per person attached and shall be:
  ◆ Designed, installed, and used under the supervision of the designated Responsible Person capable of supporting twice the weight expected to be imposed on it.
  ◆ Independent of any anchorage used to support or suspend platforms.

Positioning Device Systems. Body belt or body harness systems shall be set up so that a Soldier/employee can free fall no farther than 2 feet, and shall be secured to an anchorage capable of supporting twice the potential impact load or 3,000 pounds, whichever is greater. Requirements for snap-hooks, D-rings, and other connectors are the same as detailed in this Program under Personal Fall Arrest Systems.

Safety Monitoring Systems. In situations when no other fall protection has been implemented, the designated Responsible Person(s) shall monitor the safety of Soldiers/employees in these work areas. The designated Responsible Person(s) shall be:
- Competent in the recognition of fall hazards.
- Capable of warning workers of fall hazard dangers.
- Operating on the same walking/working surfaces as the Soldiers/employees and able to see them.
- Close enough to work operations to communicate orally with Soldiers/employees.
- Free of other job duties that might distract them from the monitoring function.

No Soldiers/employees other than those engaged in the work being performed under the Safety Monitoring System shall be allowed in the area. All Soldiers/employees under a Safety Monitoring System are required to promptly comply with the fall hazard warnings of the designated Responsible Person(s).

Safety Net Systems.
- Safety net systems must be installed no more

than 30 feet below the walking/working surface with sufficient clearance to prevent contact with the surface below, and shall be installed with sufficient vertical and horizontal distances as described in the OSHA Fall Protection Standard.
- All nets shall be inspected at least once a week for wear, damage, or deterioration by the designated Responsible Person. Defective nets shall be removed from use and replaced with acceptable nets.
- All nets shall be in compliance with mesh, mesh crossing, border rope, and connection specifications as described in the OSHA Fall Protection Standard.
- When nets are used on bridges, the potential fall area from the walking/working surface shall remain unobstructed.
- Objects that have fallen into safety nets shall be removed as soon as possible, and at least before the next working shift.

Warning Line Systems. Warning line systems consisting of supporting stanchions and ropes, wires, or chains shall be erected around all sides of roof work areas.
- Lines shall be flagged at no more than 6 foot intervals with high-visibility materials.
- The lowest point of the line (including sag) shall be between 34 and 39 inches from the walking/working surface.
- Stanchions of warning line systems shall be capable of resisting at least 16 pounds of force.
- Ropes, wires, or chains must have a minimum tensile strength of 500 pounds.
- Warning line systems shall be erected at least 6 feet from the edge, except in areas where mechanical equipment is in use. When mechanical equipment is in use, warning line systems shall be erected at least 6 feet from the parallel edge, and at least 10 feet from the perpendicular edge.

Chapter 5 –
TASKS AND WORK AREAS REQUIRING FALL PROTECTION

Facilities. Facilities shall select fall protection control measures compatible with the type of work being performed. If fall hazards cannot be eliminated, fall protection can be provided using fall protection systems and equipment. The preferred order of control measures for fall hazards are:

- Elimination - Removing the hazard from a workplace. This is the most effective control measure e.g., lower various devices or instruments, such as meters or valves to the height level of the individual, instead of servicing such devices or instruments at heights.
- Prevention (traditional) - Isolating or separating the hazard from the general work areas e.g., same level barriers such as guardrails, walls, covers or parapets.
- Work Platforms (movable or stationary) - Use scaffolds, scissor lifts or aerial lift equipment to facilitate access to work location and to protect personnel from falling when performing work at high locations.
- Personal Protective Systems and Equipment - These shall be used after other control measures (such as eliminating or isolating fall hazards) are determined not to be practical, or when a secondary system is needed e.g., when it is necessary to increase protection by employing a backup system.
- Administrative Controls - This includes introducing new work practices that reduce the risk of a person falling e.g., erecting warning lines or restricting access a work area.
- Unless otherwise specified, the designated Responsible Person(s) shall evaluate the worksite(s) and determine the specific type(s) of fall protection to be used in the following situations. A Fall Protection Program Compliance Checklist is included in Appendix C.

Excavations. Fall protection will be provided to Soldiers/employees working at the edge of an excavation that is 6 feet or deeper. Soldiers/employees in these areas are required to use the fall protection systems as designated in this program.

- Excavations that are 6 feet or deeper shall be protected by guardrail systems, fences, barricades, or covers.
- Walkways that allow Soldiers/employees to cross over an excavation that is 6 feet or deeper shall be equipped with guardrails.

Formwork and Reinforcing Steel. Fall protection will be provided when a Soldier/employee is climbing or moving at a height of over 24 feet when working with rebar assemblies.

Hoist Areas. Guardrail systems or personal fall arrest systems will be used in hoist areas when a Soldier/employee may fall 6 feet or more. If guardrail systems must be removed for hoisting, Soldiers/employees are required to use personal fall arrest systems.

Holes. Covers or guardrail systems shall be erected around holes (including skylights) that are 6 feet or more above lower levels. If covers or guardrail systems must be removed, Soldiers/employees are required to use personal fall arrest systems.

Leading Edges. Guardrail systems, safety net systems, or personal fall arrest systems shall be used when Soldiers/employees are constructing a leading edge that is 6 feet or more above lower levels. An alternative Fall Protection Plan shall be used if the designated Responsible Person(s) determines that the implementation of conventional fall protection systems is infeasible or creates a greater hazard to Soldiers/employees. All alternative Fall Protection Plans for work on leading edges shall:

- Be written specific to the particular jobsite needs.
- Include explanation of how conventional fall protection is infeasible or creates a greater hazard to Soldiers/employees.
- Explain what alternative fall protection will be used for each task.
- Be maintained in writing at the jobsite by the designated Responsible Person; and
- Meet the requirements of 29 CFR 1926.502(k).

Overhand Bricklaying and Related Work. Guardrail systems, safety net systems, personal fall arrest systems, or controlled access zones shall be provided to Soldiers/employees engaged in overhead bricklaying or related work 6 feet or more above the lower level. All Soldiers/employees reaching more than 10 inches below the walking/working surface shall be protected by guardrail systems, safety net systems, or personal fall arrest systems.

Precast Concrete Erection. Guardrail systems, safety net systems, or personal fall arrest systems shall be provided to Soldiers/employees working 6 feet or more above the lower level while erecting or grouting precast concrete members. An alternative Fall Protection Plan shall be used if the designated Responsible Person(s) determines that the implementation of conventional

fall protection systems is infeasible or creates a greater hazard to Soldiers/employees. All alternative Fall Protection Plans for precast concrete erection shall:
- Be written specific to the particular jobsite needs;
- Include explanation of how conventional fall protection is infeasible or creates a greater hazard to Soldiers/employees;
- Explain what alternative fall protection will be used for each task;
- Be maintained in writing at the jobsite by the designated Responsible Person; and
- Meet the requirements of 29 CFR 1926.502(k).

Residential Construction. Guardrail systems, safety net systems, or personal fall arrest systems shall be provided to Soldiers/employees working 6 feet or more above the lower level on residential construction projects. However, certain tasks may be performed without the use of conventional fall protection if the designated Responsible Person has determined that such fall protection is infeasible or creates greater hazards to Soldiers/employees. The designated Responsible Person shall follow the guidelines of 29 CFR 1926, Subpart M, Appendix D in the development of alternative Fall Protection Plans for construction projects (see Appendix A).

Ramps, Runways, and Other Walkways. Each Soldier/employee using ramps, runways, and other walkways will be protected from falling 6 feet (1.8 meters) or more by guardrail systems.

Roofing.
- Low-Slope Roofs. Fall protection shall be provided to Soldiers/employees engaged in roofing activities on low-slope roofs with unprotected sides and edges 6 feet or more above lower levels. The type(s) of fall protection needed shall be determined by the designated Responsible Person, and may consist of guardrail systems, safety net systems, personal fall arrest systems, or a combination of a warning line system and safety net system, warning line system and personal fall arrest system, or warning line system and safety monitoring system. On roofs 50 feet or less in width, the use of a safety monitoring system without a warning line system is permitted.

- Steep Roofs. Guardrail systems with toeboards, safety net systems, or personal fall arrest systems will be provided to Soldiers/employees working on a steep roof with unprotected sides and edges 6 feet or more above lower levels, as determined by the designated Responsible Person.

Wall Openings. Guardrail systems, safety net systems, or a personal fall arrest system will be provided to Soldiers/employees working on, at, above, or near wall openings when the outside bottom edge of the wall opening is 6 feet or more above lower levels and the inside bottom edge of the wall opening is less than 39 inches above the walking/working surface. The type of fall protection to be used will be determined by the designated Responsible Person.

Inspection and Maintenance. All fall protection equipment is inspected frequently. Visual inspections are conducted prior to each use. A Fall Protection Equipment Inspection Checklist is located in Appendix E.

Masons are the only authorized Soldiers/employees permitted to enter controlled access zones and areas from which guardrails have been removed. All other workers are prohibited from entering controlled access zones. Controlled access zones shall be defined by control lines consisting of ropes, wires, tapes, or equivalent material, with supporting stanchions, and shall be:
- Flagged with a high-visibility material at 6 feet intervals.
- Rigged and supported so that the line is between 30 and 50 inches (including sag) from the walking/working surface.
- Strong enough to sustain stress of at least 200 pounds.
- Extended along the entire length of an unprotected or leading edge.
- Parallel to the unprotected or leading edge.
- Connected on each side to a guardrail system or wall.
- Erected between 6 feet and 25 feet from an unprotected edge, except in the following cases:
  - When working with precast concrete members: between 6 feet and 60 feet from the leading edge, or half the length of the

member being erected, whichever is less.
- When performing overhand bricking or related work: between 10 feet and 15 feet from the working edge.

When guardrail systems are in use, the openings shall be small enough to prevent potential passage of falling objects. The following procedures must be followed by all Soldiers/employees to prevent hazards associated with falling objects.

- No materials (except masonry and mortar) shall be stored within 4 feet of working edges.
- Excess debris shall be removed regularly to keep work areas clear.
- During roofing work, materials and equipment shall be stored no less than 6 feet from the roof edge unless guardrails are erected at the edge.
- Stacked materials must be stable and self-supporting.
- Canopies shall be strong enough to prevent penetration by falling objects.
- Toeboards erected along the edges of overhead walking/working surfaces shall be:
  - Capable of withstanding a force of at least 50 pounds.
  - Solid with a minimum of 3 ½ inches tall and no more than 1/4 inch clearance above the walking/working surface.
- Equipment shall not be piled higher than the toeboard unless sufficient paneling or screening has been erected above the toeboard.

All Soldiers/employees who may be exposed to fall hazards are required to receive training on how to recognize such hazards, and how to minimize their exposure to them. Soldiers/employees shall receive training as soon after employment as possible, and before they are required to work in areas where fall hazards exist. A record of Soldiers/employees who have received training and training dates shall be maintained by the designated Responsible Person. A fall protection training roster can be found in Appendix H. Training of Soldiers/employees by the designated Responsible Person shall include:

- Nature of the fall hazards Soldiers/employees may be exposed to.
- Correct procedures for erecting, maintaining, disassembling, and inspecting fall protection systems.
- Use and operation of controlled access zones, guardrails, personal fall arrest systems, safety nets, warning lines, and safety monitoring systems.
- Role of each employee/Soldier in the Safety Monitoring System (if one is used).
- Limitations of the use of mechanical equipment during roofing work on low-slope roofs (if applicable).
- Correct procedures for equipment and materials handling, and storage and erection of overhead protection.
- Role of each employee/Soldier in alternative Fall Protection Plans (if used).
- Requirements of the OSHA Fall Protection Standard, 29 CFR 1926, Subpart M.
- The organization requirements for reporting incidents that causes injury to a Soldier/employee.
- Hands-on training is also required for investigation and inspection work.
  - Personnel exposed to fall hazards shall receive refresher/update training on the safe use of fall protection equipment and rescue in the following:
  - End-users, program manager/administrator, and the authorized rescuer (in-house person who conducts rescue): Fall protection - minimum of one-hour refresher training annually to stay current with fall protection and rescue requirements.
  - Fall Protection Competent and Qualified Person for Fall Protection: Refresher/update training every two years.
  - Additional training shall be provided on an annual basis, or as needed when changes are made to this Fall Protection Program, an alternative Fall Protection Plan, or the OSHA Fall Protection Standard.

All incidents that result in injury to workers, as well as near misses, regardless of their nature, shall be reported and investigated IAW AR 385-10. Investigations shall be conducted by the designated Responsible Person as soon after an incident as possible to identify the cause and means of prevention to eliminate the risk of reoccurrence.

In the event of such an incident, the Fall Protection Program (and alternative Fall Protection Plans, if in place) shall be reevaluated by the designated Responsible Person to determine if additional practices, procedures, or training is necessary to prevent similar future incidents.

Chapter 10 –
CHANGES TO THE PLAN

Any changes to the Fall Protection Program (and alternative Fall Protection Plans, if in place) shall be approved by the designated Responsible Person, and shall be reviewed by a Qualified Person as the job progresses to determine additional practices, procedures or training needs necessary to prevent fall injuries. Affected Soldiers/ employees shall be notified of all procedure changes, and trained if necessary. A copy of this plan, and any additional alternative Fall Protection Plans, shall be maintained at the jobsite by the designated Responsible Person.

Chapter 11 –
INSPECTION, STORAGE, MAINTENANCE AND CARE PROCEDURES FOR PERSONAL FALL PROTECTION EQUIPMENT

29 CFR 1910 and 1926 state that personal fall arrest systems must be regularly inspected. Any component of the system with significant defects must be removed from service immediately and shall be tagged or marked as unusable or destroyed. All fall protection equipment shall be inspected before each use by the user in accordance with the manufacturer's instructions. Prior to each use inspection means that the fall protection equipment shall be inspected by the end-user at least once at the beginning of each eight-hour shift in which it is used to verify that it has not sustained any wear or damage that would require its removal from service. The Competent Person shall inspect the equipment at intervals of no more than one year or as prescribed by the manufacturer of the equipment. Most manufacturers recommend inspection of the equipment to be conducted twice annually by the Competent Person. Inspection of the equipment by the Competent Person shall be documented and the tag on the equipment shall be checked and dated by the Competent Person on the date of inspection. All components and sub-components of the selected fall-arrest, positioning, and restraint systems shall be compatible. As a general rule, always consult equipment manufacturers' instructions and recommendations for use, inspection, care and maintenance procedures. A Fall Protection Equipment Inspection Checklist is included in Appendix E.

- Inspect personal fall protection equipment for the following defects:
  - Webbing and ropes (harnesses, lanyards, straps, etc.): Look for cuts, wear, tears, damaged threads, broken fibers, undue stretching, torn or pulled stitches, frayed edges, mold, alterations, or additions which will affect its efficiency, damage due to deterioration, chemical damage (contact with fire, acids, or corrosives), abrasions, ultraviolet deterioration and missing markings and labels, and any wearing or internal deterioration of the ropes.
  - Hardware (snap-hooks, carabiners, connectors and D-rings): Look for distorted hooks or faulty springs, tongues unfitted to the shoulder buckles, loose or damaged mountings, nonfunctional parts; check for signs of excessive wear, crack, corrosion and deformation.

- Anchorage Systems (anchorages and anchorage connectors):
  - Inspect all components of the anchorage systems.
  - Observe any abrasions, wear points, damaged threads, or sags in the sling material before use.
  - For synthetic slings and anchor straps inspect all sewing and loops for wear, chemical damage, burn damage, and/or ultraviolet deterioration.
  - Refer to the anchorage-attached tags to determine when the sling should be retired.
  - Inspect cable slings for excessive damage to the steel fibers.
  - Re-certify the anchorage system exposed to weather or corrosive conditions.
  - Inspect anchorage connectors for integrity and attachment to solid surfaces.

- Snap-Hooks and Carabiners
  - Inspect on a regular basis and before each use.
  - Inspect snap-hooks and carabiners for any hook, locks and eye distortion.
  - Verify there are no cracks, pitted surfaces and eye distortions.
  - The keeper latch must not be bent, distorted, or obstructed.
  - Verify that the keeper latch seats into the nose without binding.

- Verify that the keeper spring securely closes the keeper latch.
- Test the locking mechanism to verify that the keeper latch locks properly.
- Verify the points where the lanyard attaches to the snap-hooks are free of defects.
- Retire snap-hooks, carabiners, and all integral components if any discoloration, deformation, cracks, or abrasions are detected.
- Retire immediately if the item has sustained any fall, or if the spring is broken and the gate is bent, or if the gatekeeper no longer engages the slot cleanly.
- Damaged snap-hook and carabiners shall be tagged and removed from service and from the inventory list.
- Dirty snap-hooks and carabiners shall be cleaned with kerosene, WD-40 (or equivalent), or similar solvents, and immersed in boiling water for 30 seconds to remove the cleaning agent. Dry with a soft cloth to ensure that the gate and gatekeeper operate properly.
- Ensure that only double-locking-type gates are used.

• Lanyards and Energy Absorbers
- Inspect lanyards put under a slight tension on a regular basis.
- Check all components for abrasion, cuts, discoloration, cracks, burns, knots, torn stitching and excessive wear.
- Visually inspect the energy absorber for any signs of damage, paying close attention to where the energy absorber attaches to the lanyard.
- Wash lanyards and energy absorbers on a regular basis to remove dirt and grit, which can abrade the fibers.
- Lanyards and energy absorbers shall have permanently attached labels indicating the manufacturer's name, serial number or lot number, date of manufacture, maximum elongation (deployment distance), maximum and average arresting force, maximum free-fall distance (6 or 12-foot free fall), and capacity. The lanyards and energy absorbers must alsohave permanently attached labels that indicate they meet OSHA & ANSI Z359.13 Standard and requirements. Lanyards bearing the markings of ANSI A10.14, ANSI Z359.1 (1992, R199) or ANSI Z359.1 (2007) are not acceptable and they shall be taken out of service.
- Use and review the manufacturer's logbook provided with the equipment to determine the age of the lanyard and energy absorber.
- Lanyards and energy absorbers shall be inspected by the user prior to each use and by a Competent Person other than the user at least once a year.
- Check for missing marking and labels.
- Maximum usage of a lanyard shall not be more than five years, unless the Competent Person for fall protection carefully inspects it, review its history of use and storage, and recommends its continued use, once put in service (assuming the new unused lanyard is stored in a climate-controlled location, [i.e., in a plastic bag not exposed to fumes, and in a cool location out of direct sunlight]).
- Retire the lanyard:
  ♦ After a hard fall.
  ♦ When the shock absorber (even if slightly) is impacted or deployed.
  ♦ If the lanyard has been used for any other purpose other than Fall Protection.
  ♦ If the equipment shows excessive wear, chemical damage, burn damage, and/or ultraviolet deterioration.

• Fall-Arrestor (Rope Grab)
- Inspect regularly.
- Check for signs of wear, corrosion, rust, and other anomalies.
- If any sign of wear or malfunction is observed, remove the device from service immediately.

• Self-Retracting Devices
- Inspect before each use for any physical damage.
- Inspection by a Competent Person shall be in accordance with the manufacturer's instructions and recommendations. The inspection shall be documented.
- If the SRD housing becomes yellow, gathers condensation, or the indicator has been engaged, remove it from service immediately, and return it to the manufacturer for repair and re-certification.
- SRDs shall have permanently attached labels indicating that they meet ANSI Z359.14 and OSHA Standards and requirements.
- Make sure all back nuts or rivets are tight
- Make sure the entire length of the nylon strap is free of any cuts, burns, abrasions, kinks, knots, broken stitches, and excessive wear, and retracts freely.
- Test the unit by pulling sharply on the lanyard to verify the locking mechanism is operating correctly.

Additional Discussion. SRDs should be briefly inspected prior to each use, and more thoroughly inspected by a

Competent Person regularly. With specialized training, it is possible that a Competent Person can become certified to conduct re-certification and general services. Usually, SRDs are returned to the manufacturer for service and recertification. Any equipment with many movable mechanical components or parts requires specialized inspection. Usually the Competent Person does not have the tools, equipment and/or qualification to conduct such inspection. In order to determine if the SRD is in good and safe working condition, specialized testing and inspection must be conducted on the SRD. This includes opening the casing, inspecting the inner components of the SRL and the drum containing excess spooled line, inspecting the locking mechanism, spring, connecting means, and fall indicator, and corrosion inspection in special environments. This is the why only the manufacturer can inspect and certify the SRD.

SRDs are designated as Repairable or Non-repairable. Repairable SRDs shall be returned to the manufacturer for servicing and re-certification (factory-authorized inspection) depending on the type, usage and the environment, in accordance with the following Inspection Requirements table: All factory authorized inspections of all self-retracting devices shall not be longer than the intervals required by the manufacturer.

SRDs can either be repaired by the manufacturer or cannot be repaired. For repairable SRDs, not requiring factory-authorized inspections by a manufacturer would be considered non-compliant with ANSI Z359 standards.

TYPE OF USE	APPLICATION EXAMPLES	CONDITION OF USE	INSPECTION FREQUENCY BY COMPETENT PERSON	FACTORY AUTHORIZED INSPECTION
Infrequent to light use	Used in rescue, confined space, industrial (factory) maintenance	Good storage conditions, indoor or infrequent outdoor use, room temperature, clean environments	Annually	At least every 2-5 years, but no longer than interval required by the manufacturer
Moderate to heavy use	Transportation Facilities, Residential/wood Construction, Utilities and Warehouses/Hangars	Fair storage conditions, indoor and extended outdoor use, all temperature, clean or dusty environments	Semi Annually To annually	At least every 1-2 years, but no longer than interval required by the manufacturer
Continuous usage to severe conditions	Commercial, Construction and industrial use, Shipyard environment	Harsh storage conditions, prolonged or continuous outdoor use, all temperatures, dirty environment	Quarterly to semi annually	At least annually, but no longer than intervals required by the manufacturer

Non-repairable SRDs are designed and designated as non-repairable (not designed for disassembly). They are very basic in design and limited in length and use, made of synthetic webbing. Complete internal inspection is not possible without destroying or damaging the device. The Competent Person for fall protection shall work with written manufacturer's inspection requirements to determine whether the device can be used or not. If the inspection is not possible, the device shall be taken out of service. The contrast of the device being used inside or outside would affect the term of its use dramatically.

- Body Support (Full-Body Harness)
  - Inspect daily or before each use.
  - Inspect thoroughly and verify there are no torn, frayed, broken fibers, pulled stitches, or frayed edges, anywhere on the harness
  - Closely examine all of the nylon webbing to ensure there are no burn marks from welding or heat sources, which could weaken the material
  - Examine D-ring for excessive wear, deterioration, or cracks.
  - Verify buckles are not deformed, cracked, and will operate correctly.
  - Check to see that all grommets (if present) are secure and not deformed from abuse or a fall.
  - Check tongue/straps for excessive wear from repeated buckling.
  - All rivets must be tight, not deformed.
  - Inspect for missing markings and labels.
  - Ensure that harnesses are not painted or marked.
  - Examine the harness for discoloration, abrasions and ultraviolet deterioration.
  - Store harnesses in a cool, dry, and safe environment, ideally in a locked storage area.
  - A Competent Person other than the user shall inspect the harness periodically, or at least once a year.
  - Wash the harness in a mild soap and rinse multiple times to remove any soap residue, and hang to dry out of direct sunlight in a cool, dry environment.
  - Maintain a logbook indicating the date of entry into service, the nature of the work performed, washing of the harness, or other relevant details.
  - Retire harnesses from service after five years from the date put into service (assuming the new unused harness is stored in a climate-controlled environment [i.e., in a plastic bag not exposed to fumes, and in a cool location out of direct sunlight]) unless the Competent Person for fall protection carefully inspects it, reviews its history of use and storage, and recommends its continued use.
  - The body support harness shall have a permanently attached label indicating manufacturer's name, serial number or lot number, date of manufacture, capacity, and that it meets OSHA & ANSI Z359 Fall Protection Code/Standards requirements.

- Ropes (Synthetic Fibers)
  - Inspect rope periodically for broken fibers, severely worn areas, or change in the consistency of the core; inspect under slight tension and check for soft areas, bulges, or excessive stiffness.
  - Avoid exposing rope to hazardous chemicals, moisture, acids, or oils.
  - Do not use the rope after it is impacted or damaged.
  - Wash the rope on a regular basis with lukewarm water and mild detergent to remove dirt or grit, rinse several times to remove soap residue, and hang in a dry, cool, dark area.
  - Store rope in a strong weatherproof bag. Rope always must be dry before being placed in storage.
  - Rope shall have a permanently attached label indicating manufacturer's name, serial number or lot number, date of manufacture, capacity, and that it meets OSHA & ANSI Z359 Fall Protection Code/Standards requirements.
  - Retire rope after five years of service unless the Competent Person for fall protection carefully inspects it, reviews its history of use and storage, and recommends its continued use. If it is damaged, impacted, or exposed to chemicals, remove from service immediately.
  - Avoid the use of Kernomantle ropes.

- Single Anchor Vertical Lifelines (Flexible). Refer to Ropes (synthetic fibers) above and manufacturer's recommendations regarding inspection, care, and maintenance.

- Climbing Ladder Fall Arrest System
  - Inspect on a regular basis per equipment manufacturer's requirements.
  - The sleeve must run freely without hand operations or guidance.
  - Check cable and rails for abrasions, wear, looseness, and cracks.
  - Before climbing, check integrity of cable, system, and ground level.

- Raising/Lowering Devices (Rescue)
  - Inspect before each use.
  - Check for wear and corrosion.

- Horizontal Lifeline (HLL)
  - Inspect the system including anchorages, anchorage connectors, cable and other hardware for defects or loose or components similar to inspection of other fall-arrest system components. The end-user shall inspect the components of the system prior to each use.

- Type 1 HLL system, the Competent Person for fall
  protection shall inspect the system at an interval
  of no more than one year under the direction
  of a Qualified Person for fall protection.
- Type 2 HLL System shall be inspected once a year
  by the Competent Person for fall protection who
  is trained by the manufacturer of the system to
  perform such inspections and under the direction
  of the Qualified Person for fall protection.

- Additional Instructions for Assembly, Disassembly,
  Storage Inspection, Care and Maintenance
  - Protect against cuts and abrasions: All safety lines and
    lanyards shall be protected against cuts or abrasions.
    Padding must be used wherever sharp edges exist.
  - All safety lines and lanyards shall be stored
    in an approved location, as follows:

- All personal fall protection equipment shall be stored
  in a weatherproof container or locker when not in
  use. Equipment should not be allowed to lie in water
  or direct sunlight, since this will affect equipment
  strength. Never store personal fall-arrest equipment
  in the bottom of a toolbox, on the ground, or outside
  exposed to the elements (i.e., sun, rain, snow, etc.).
  - The fall-arrest system components shall be compatible,
    as follows: Contact the qualified or Competent
    Persons for fall protection or manufacturer's
    representative for assistance. In the use of fall-
    arrest systems, all components shall be designed for
    use with each other, or approval must be obtained
    from the manufacturer or Qualified Person to use
    the configuration that uses different components.
    All system components shall be compatible.
  - Follow the manufacturers and the Qualified Person's
    instructions for installation, assembly/disassembly,
    and use, as follows: All systems must be installed,
    assembled, and disassembled per the manufacturer's
    direction. Failure to follow these instructions
    could lead to the possible failure of a system.
  - In the event of a fall, secure all equipment involved
    and contact the Safety Officer for disposition. Do
    not reuse safety equipment that has experienced
    a fall: In the event of a fall, the first response is to
    ensure the safety of the Soldiers/employees. After
    rescue and, if required, medical aid is provided, all
    equipment involved must be removed from service.
  - During inspection of the equipment by the Competent
    Person for fall protection, only mark on the labels.
    Some manufacturers permit marking on straps
    using certain types of permanent markers which are
    water resistant and quick-drying such as Sanford
    Sharpie® Permanent Markers. Always consult the
    manufacturer for marking on the equipment.
  - Care and maintenance of the equipment.
    Always consult manufacturer's instructions
    and recommendations for care and
    maintenance of the equipment.
    - Snap-hooks and Carabiners: Clean dirty gates of
      snap-hooks and carabiners by applying WD-40
      (or equivalent), other solvents, oil, or kerosene,
      until the gates work smoothly, than immerse
      in boiling water for 20-30 seconds to remove
      cleaning agent; dry with a soft cloth to ensure
      that the gate and gatekeeper operate properly.
    - Harnesses, Lanyards and Ropes: Wash on regular
      basis with mild soap and rinse multiple times to
      remove the soap residue, store in a cool dry and safe
      environment to dry. Ensure harnesses and lanyards
      are not painted or marked. Mark only on labels.

Chapter 12 –
AVIATION FALL PROTECTION

Introduction. Falls from aircraft are potential sources of
injuries and fatalities to aircraft maintenance personnel,
aircrew, and inspectors. Specific guidance to mitigate these
risks and help to ensure the safety of all personnel who
perform aviation maintenance and inspection work at the
organizational, intermediate, and depot levels is required.

Applicability. This guidance document applies to all
Army Military and Civilian personnel worldwide involved
in aircraft maintenance and inspection work where
personnel are exposed to the hazard of falling from
heights and/or there is a need for fall protection.

Purpose. The purpose of this Chapter is to provide
administrative tools, criteria, and safe work
practices to mitigate fall hazards when conducting
aircraft maintenance or inspections.

Applicable Standards, Regulations, SOPs and Instructions. 29 CFR 1910 requires fall protection when working 4 feet or more above lower levels to be mitigated by one or more of five hierarchies of control measures. In addition, federal OSHA fall protection regulations apply to civilian personnel at all times. Federal OSHA regulations 29 CFR 1910 also apply to military personnel conducting sustainment level maintenance at all times, and field level maintenance when conducting maintenance or inspections not directly involved in military flight operations. The use of fall protection shall be used to the maximum extent feasible, and the use of deliberate risk management is required for all aviation maintenance and/or inspection evolutions. This written program must be site specific and tailored to the individual command's equipment and requirements based on the hazard assessment. For example, if the unit does not utilize harness type fall protection equipment, the associated items such as rescue procedures may be omitted. Alternate type/model/series specific templates should be developed for your particular aircraft. The command policy should contain two appendices. Appendix one should be the Fall Hazard Survey and appendix two should be the Fall Protection and Prevention Plan. Although Appendix A has a sample written Fall Protection Plan, this appendix includes a template more specific to aviation that commands can tailor to their needs.

Duties and Responsibilities. The command Fall Protection Program Manager (FPPM) shall ensure that assigned personnel have the necessary skills, knowledge, training, and expertise to manage, administer, and implement the fall protection program. At a minimum, the command shall have a designated FPPM, and end-users of fall protection. The fall protection program provides additional guidance for positions within the program, including the duties and responsibilities of personnel involved in fall protection. The following positions shall be included in the command fall protection program with their required qualifications and responsibilities:

- Aviation Fall Protection Program Manager (Aviation FPPM). A person designated in writing by the Commanding Officer who is responsible for the development and implementation of the Fall Protection Program. The Aviation FPPM shall ensure that personnel exposed to fall hazards and other personnel involved in the program receive adequate training. Aviation FPPMs may contact their Safety Office for information on possible Aviation FPPM training options. Designation of Aviation (FPPM) to oversee the administrative requirements, equipment selection, and inspection aspects of the program— this responsibility should be assigned to the command Safety Officer or appropriately trained individual.

- Qualified Person for Fall Protection (optional). A person with a recognized engineering degree or professional certificate and with extensive knowledge, training, and experience in fall protection and rescue field; who is capable of performing design, analysis, and evaluation of fall protection systems and equipment.

- Competent Person for fall protection. A person designated in writing by the Commanding Officer to be responsible for the immediate supervision, implementation and monitoring of the fall protection program, who is capable of identifying, evaluating and addressing existing and potential fall hazards, and in the application and use of personal fall protection and rescue system or any component thereof, and has the authority to take prompt corrective measures to eliminate or control the hazards of falling. To be trained to the Competent Person level, personnel must attend a Competent Person training course.

- End-User for Fall Protection Equipment (Authorized Person) – Shall be trained in the use of personal fall protection equipment

- Competent Person for inspection of fall protection equipment (as required)—A person who has been trained by the Competent Person or Qualified Person for fall protection to perform and document inspections of personal fall protection equipment. Personnel trained as an Aviation FPPM can also perform these inspections.

- Exposed Personnel. Any person exposed to a fall hazard while performing aircraft maintenance and inspections. Exposed personnel shall receive unit level fall protection training from the Aviation FPPM. At a minimum training shall include: stands, ladders, over-wing maintenance and climbing aircraft.

Workplace Survey and Assessment of Fall Hazards
- Workplace Survey – A Competent Person for Fall Protection or an Aviation FPPM shall conduct the workplace survey. Appendix J provides a template for conducting a workplace survey on a UH-60. The workplace survey shall encompass the aircraft and all maintenance areas (i.e. hangars,

wash racks, flight line) and different access equipment that may be available in each area. The fall hazard survey shall be validated annually for effectiveness.

- Hazard Assessment – Once fall hazards are identified in the workplace survey, the hazards must be assessed. The template for conducting aviation workplace surveys provides matrices to assess hazards based on mishap probability and severity. Considerations such as potential environmental conditions should be included in each hazard assessment.
- Workplace Survey Report – Identification and assessment of fall hazards in addition to comprehension of the tasks to be performed by personnel working at heights which will allow the Competent Person for fall protection or program manager to develop alternatives to mitigate fall hazards. Appendix B of the Fall Protection Program Guide provides guidance for preparing workplace survey reports. The Workplace Survey Report should list all fall hazards (height and hazards for each elevated working platform or area), provide a list of maintenance tasks that may be executed for each elevated working area, and provide a risk assessment for each task. Tasks may be grouped for assessment purposes for each elevated working area if they pose the same risk. Caution must be applied to grouping tasks that have different risk assessments. For example, applying torque to a main rotor head bolt may present a much greater hazard severity and probability than inspection of the main rotor head servicing despite both tasks being conducted on the same elevated working platform. A Workplace Survey Report is site specific, but reports for a particular type/model/series may have already been completed and can serve as a template for alternate sites. Contact your Safety Officer for potential templates.

Aircraft Fall Hazard Prevention and Controls. Each task identified in the workplace survey report must be mitigated with one of the five control methods listed in Chapter 6 according to the hierarchy or preferred order of control measures for fall hazards. Each maintenance task (or tasks as grouped in the workplace hazard assessment) to be executed and its corresponding control method(s) must be delineated as part of the site specific Fall Protection and Prevention Plan. This plan is to be included in the organizations fall protection program instruction. Contact your Safety Officer to determine if a fall protection and prevention plan has already been developed for your airframe. Additional general guidance to be included in fall protection and prevention plans, as applicable, follows:

- Aircraft Maintenance and Inspection Work
  - Personnel shall be trained to recognize the hazards of falling, fall risks at the worksite, recognition of fall-hazard deficiencies, and safe use of the equipment they are operating on including integrated features such as steps and hold points, and the selection and safe use of fall protection equipment.
  - Designated walkways shall be identified and used wherever possible. Walkways are pre-identified paths that a person is permitted or allowed to walk on without the use of fall protection equipment.
  - Personnel working on aircraft surfaces should wear slip-resistant soled shoes, head protection, and other appropriate PPE.
  - Good housekeeping practice is paramount, and shall be enforced or implemented.
  - Clean all aircraft surfaces immediately when hydraulic fluids, oils, and other fluids contaminate the worksite.
  - Winds or other environmental variables such as rain, snow, frost, or ice that may preclude the safe performance of maintenance or inspection work shall be considered in the RM executed for that evolution.
  - Where the use of fall protection equipment is not feasible (e.g., active flight line, preflight inspections); commands shall utilize RM, at a minimum, to analyze and determine alternate methods (administrative controls) to mitigate risk.

- Cleaning, Washing or Deicing Aircraft
  - To protect against falls while cleaning, washing, or deicing aircraft, personnel should not be allowed to climb or walk on wet, frost-covered, or icy surfaces.
  - For washing or cleaning aircraft, separate elevated work platforms or work stands and long-handle brushes shall be used to the maximum extent possible. This does not preclude the additional use of other fall protection systems.

- Where it is necessary to walk on aircraft wings or other surfaces during washing, extreme care shall be exercised and other control measures, as appropriate, shall be utilized, such as horizontal life-lines or self-retracting-lanyards (SRLs), to which a full-body harness can be attached.

- Fall protection systems and equipment used for aircraft maintenance and inspection work.

Fall Protection and Fall Protection Methodologies. Fall protection and fall protection methodologies are pieces of the overall hazard analysis and prevention plan for an aviation maintenance and/or inspection evolution. Location of the aircraft or potential fall exposure, nature of the task, environmental conditions, work area of the aircraft or working platform, and consideration for other potential hazards that may be introduced with the use of fall protection methodologies, shall be considered for each task. Consideration must be given to all hazards encountered with the execution of a particular maintenance or inspection task, to determine the best and safest course of action. For example, moving aircraft in order to make space for fall protection equipment, especially in confined and/or areas with high density of other aircraft and equipment, adds additional hazards that may outweigh those of the fall hazard. Special consideration must be given to the use of fall protection equipment and systems on active flight lines due to the risk of foreign object damage (FOD) and the potential effects of rotor, jet, or propeller wash. If maintenance or inspection work is performed on the flight line, consideration should be given to utilizing a designated maintenance area that is free from the additional hazards that may be caused by propeller, jet, or rotor wash. Where these hazards cannot be controlled effectively, then administrative controls may provide the best fall protection methods on active flight lines. In this case, personnel shall use deliberate risk management and other administrative controls to mitigate fall hazards. If the Competent Person for fall protection determines that the use of fall protection equipment is not feasible or practicable based on the overall hazard analysis, personnel shall use deliberate risk management and other administrative controls to mitigate fall hazards in accordance with guidance of the Competent Person for fall protection. A primary objective of the Fall Protection and Prevention Plan, is to provide site specific, standard methods to mitigate fall hazards associated with routine tasks and to provide guidance, should deviation to standard methods be required. The following paragraphs list types of fall protection systems that can be used for aircraft maintenance and inspection, in cases where fall hazards cannot be eliminated. Units must determine those fall protection measures which must be taken, based on risk analysis of the work or inspection to be performed.

Mobile Work Platforms (MWP). Where work is performed from elevated work platforms four feet or higher, the work platforms shall be equipped with a standard guardrail and safe method of access, or other fall protection system that mitigates potential fall hazards. Mobile servicing platforms are authorized, but shall be required to provide additional fall protection equipment per the manufacturer's specifications. MWPs provide powered access from self-propelled elevating platforms which are approved as Ground Support Equipment.

Other Aerial Work Platforms
- B-Stands
- Phase Stands
- Warehouse Stands
- Ladders

Restraint System. A system consisting of equipment and components connected together designed to restrain a person from reaching an exposed fall hazard.

Personal Fall-Arrest System. A system used to arrest a person during a fall from a working level. It consists of an anchorage system, connecting means, and full-body harness and may include a lanyard, self-retracting device, lifeline, or suitable combination of these. A personal fall arrest system shall be rigged so that Soldier/employee arrested fall will not come in contact with a lower level or object. Safety belts (body belts) shall not be used. Suitable anchor points for personal fall arrest systems include horizontal lifeline systems, moveable and/or fixed anchorages attached to a rigid rail or beam, and vacuum type anchors. Vacuum anchors shall not be used on composite or thin shelled aircrafts. They shall only be used on thick shelled airframes.

Administrative Controls. These are controls that reduce risks through specific administrative actions. Methods for implementing administrative controls include:

- Fall Hazard Awareness Training for personnel performing elevated work.
- Providing suitable warnings, markings, placards, signs and notices.
- Establishing written policies, programs, instructions, and SOPs.
- Conducting a deliberate RM with consideration of the following as a minimum.
  - Maintenance or inspection work duration,
  - Environmental factors such as wind, rain, snow, or ice,
  - Probability of fall,
  - Hazards of fall (obstacles, height, etc.)
  - Slippery materials or substances on aircraft surfaces
  - Actual requirement for work to be completed
- Limiting the exposure to a hazard (by reducing the number of personnel, and/or the length of time personnel are exposed).

Inspection, Care, Maintenance and Storage of Fall Protection Equipment. The Fall Protection Program Guide provides guidance, checklists, and specific requirements for this topic. The fall protection program instruction must address specific inspection, maintenance, storage, and care procedures for the fall protection equipment possessed and/or utilized by the command. Manufacturers' instructions and recommendations may provide a good starting point for these requirements. Checklists, maintenance requirement cards, and pre-operational inspection cards should be developed for each piece of fall protection equipment by a Competent Person for fall protection or fall protection program manager and approved by the responsible authority.

Rescue Procedures. The Fall Protection Program Guide provides guidance and templates for this requirement. A site-specific rescue plan shall be prepared in writing and maintained for all instances where personnel work at heights while utilizing harnessotype fall protection systems. If the rescue will be performed by the fire department or other governmental jurisdictional agency, a written pre-incident plan is required. The rescue plan shall contain detailed procedures on the methods of rescue to include methods of self-rescue/assisted rescue, equipment used in the rescue procedure, training requirements including specialized training for rescuers, procedures for requesting rescue, and a pre-mishap medical plan should medical assistance be required. The rescue plan should be included as part of the written Fall Protection and Prevention Plan.

Audits and Evaluations. The Fall Protection Program Guide provides compliance checklist for the Fall Protection Program which can be used for auditing the program. Program audits should be conducted semiannually, but shall be conducted annually as part of the command safety self-assessment.

Examples of Fall Protection Program Documents Templates. The following are templates to assist units to establish, implement and manage viable fall protection programs. These templates include the following:
- Written Fall Protection Program Template
- Fall Hazard Survey Report Template
- Fall Protection and Prevention Plan Template

Chapter 13 – GROUND MAINTENANCE FALL PROTECTION

Introduction. Falls from vehicles are potential sources of injuries and fatalities to vehicle maintenance personnel, aircrew, and inspectors. Specific guidance to mitigate these risks and help to ensure the safety of all personnel who perform vehicle maintenance and inspection work at the organizational, intermediate, and depot levels is required.

Applicability. This guidance document applies to all Army Military and Civilian personnel worldwide involved in vehicle maintenance and inspection work where personnel are exposed to the hazard of falling from heights and/or there is a need for fall protection.

Purpose. The purpose of this chapter is to provide administrative tools, criteria, and safe work practices to mitigate fall hazards when conducting vehicle maintenance or inspections.

Applicable Standards, Regulations, SOPs and Instructions. 29 CFR 1910 requires fall protection when working 4 feet or more above lower levels to be mitigated by one or more of five hierarchies of control measures. In addition, federal OSHA fall protection regulations apply to civilian personnel at all times. Federal OSHA regulations 29 CFR 1910 also apply to military personnel conducting intermediate level and depot level maintenance at all times, and organizational level maintenance when conducting maintenance or inspections not directly involved in military operations. The use of fall protection shall be used to the maximum extent feasible, and the use of deliberate risk management is required for all vehicle maintenance and/or inspection evolutions. This written program must be site specific and tailored to the individual command's equipment and requirements based on the hazard assessment. For example, if the unit does not utilize harness type fall protection equipment, the associated items such as rescue procedures may be omitted. Alternate type/model/series specific templates should be developed for your particular vehicle. The command policy should contain two appendices. Appendix one should be the Fall Hazard Survey and appendix two should be the Fall Protection and Prevention Plan. Although Appendix A of this document has a sample written construction Fall Protection plan, this appendix includes a template more specific to vehicle maintenance that commands can tailor to their needs.

Duties and Responsibilities. The command FPPM shall ensure that assigned personnel have the necessary skills, knowledge, training, and expertise to manage, administer, and implement the fall protection program. At a minimum, the command shall have a designated FPPM, and end-users of fall protection. The fall protection program provides additional guidance for positions within the program, including the duties and responsibilities of personnel involved in fall protection. The following positions shall be included in the command fall protection program with their required qualifications and responsibilities:

- Unit Fall Protection Program Manager (FPPM). A person designated in writing by the Commanding Officer who is responsible for the development and implementation of the Fall Protection Program. The FPPM shall ensure that personnel exposed to fall hazards and other personnel involved in the program receive adequate training. FPPMs may contact their Safety Office for information on possible training options. Designation of the FPPM to oversee the administrative requirements, equipment selection, and inspection aspects of the program—this responsibility should be assigned to the command

Safety Officer, or appropriately trained individual.

- Qualified Person for Fall Protection (optional). A person with a recognized engineering degree or professional certificate and with extensive knowledge, training, and experience in fall protection and rescue field; who is capable of performing design, analysis, and evaluation of fall protection systems and equipment.
- Competent Person for Fall Protection. A person designated in writing by the Commanding Officer to be responsible for the immediate supervision, implementation and monitoring of the fall protection program, who is capable of identifying, evaluating and addressing existing and potential fall hazards, and in the application and use of personal fall protection and rescue system or any component thereof, and has the authority to take prompt corrective measures to eliminate or control the hazards of falling. To be trained to the Competent Person level, personnel must attend a Competent Person training course.
- End-user for Fall Protection Equipment (Authorized Person) – Shall be trained in the use of personal fall protection equipment
- Competent Person for Inspection of Fall Protection Equipment (as required) — A person who has been trained by the Competent Person or Qualified Person for fall protection to perform and document inspections of personal fall protection equipment. Personnel trained as a Unit Fall Protection Program Manager can alsoperform these inspections.
- Exposed Personnel. Any person exposed to a fall hazard while performing vehicle maintenance and inspections. Exposed personnel shall receive unit level fall protection training from the FPPM. At a minimum training shall include: stands, ladders, and climbing vehicles.

Workplace Survey and Assessment of Fall Hazards

- Workplace Survey – A Competent Person for Fall Protection or a FPPM shall conduct the workplace survey. Appendix B provides a template for conducting workplace surveys. The workplace survey shall encompass the vehicle and all maintenance areas (i.e. hangars, wash racks, parking area) and different access equipment that may be available in each area. The fall hazard survey shall be validated annually for comparison purposes.
- Hazard Assessment – Once fall hazards are identified in the workplace survey, the hazards must be

assessed. The template for conducting workplace surveys provides matrices to assess hazards based on mishap probability and severity. Considerations such as potential environmental conditions should be included in each hazard assessment.

- Workplace Survey Report – Identification and assessment of fall hazards in addition to comprehension of the tasks to be performed by personnel working at heights which will allow the Competent Person for fall protection or program manager to develop alternatives to mitigate fall hazards. Appendix B of the Fall Protection Program Guide provides guidance for preparing workplace survey reports. The Workplace Survey Report should list all fall hazards (height and hazards for each elevated working platform or area), provide a list of maintenance tasks that may be executed for each elevated working area, and provide a risk assessment for each task. Tasks may be grouped for assessment purposes for each elevated working area if they pose the same risk. Caution must be applied to grouping tasks that have different risk assessments. A Workplace Survey Report is site specific, but reports for a particular type/model/series may have already been completed and can serve as a template for alternate sites. Contact your Safety Officer for potential templates.

Fall Hazard Prevention and Controls. Each task identified in the workplace survey report must be mitigated with one of the five control methods listed in Chapter 6 according to the hierarchy or preferred order of control measures for fall hazards. Each maintenance task (or tasks as grouped in the workplace hazard assessment) to be executed and its corresponding control method(s) must be delineated as part of the site specific Fall Protection and Prevention Plan. This plan is to be included in the unit fall protection program instruction. Contact your Safety Officer to determine if a fall protection and prevention plan have already been developed for your vehicle. Additional general guidance to be included in fall protection and prevention plans, as applicable, follows:

- Maintenance and Inspection Work
  - Personnel shall be trained to recognize the hazards of falling, fall risks at the worksite, recognition of fall-hazard deficiencies, and safe use of the equipment they are operating on including integrated features such as steps and hold points, and the selection and safe use of fall protection equipment.
  - Designated walkways shall be identified and used wherever possible. Walkways are pre-identified paths that a person is permitted or allowed to walk on without the use of fall protection equipment.
  - Personnel working on vehicle surfaces should wear slip-resistant soled shoes, head protection, and other appropriate PPE.
  - Good housekeeping practice is paramount, and shall be enforced or implemented.
  - Clean all vehicle surfaces immediately when hydraulic fluids, oils, and other fluids contaminate the worksite.
  - Winds or other environmental variables such as rain, snow, frost, or ice that may preclude the safe performance of maintenance or inspection work shall be considered in the RM executed for that evolution.
  - Where the use of fall protection equipment is not feasible (e.g., operations); commands shall utilize RM, at a minimum, to analyze and determine alternate methods (administrative controls) to mitigate risk.

- Cleaning, Washing or Deicing Vehicle
  - To protect against falls while cleaning, washing, or deicing vehicles, personnel should not be allowed to climb or walk on wet, frost-covered, or icy surfaces.
  - For washing or cleaning vehicles, separate elevated work platforms or work stands and long-handle brushes shall be used to the maximum extent possible. This does not preclude the additional use of other fall protection systems.
  - Where it is necessary to walk on vehicle surfaces during washing, extreme care shall be exercised and other control measures, as appropriate, shall be utilized, such as horizontal life-lines or self-retracting-lanyards (SRLs), to which a full-body harness can be attached.

Fall Protection Systems and Equipment Used for Vehicle Maintenance and Inspection Work. Fall protection and fall protection methodologies are pieces of the overall hazard analysis and prevention plan for a vehicle maintenance and/or inspection evolution. Location of the vehicle or potential fall exposure, nature of the task, environmental conditions, work area of the vehicle or working platform, and consideration for other potential hazards that may be introduced with the use of fall protection methodologies, shall be considered for each task. Consideration must be given to all hazards encountered with the execution of a particular maintenance or inspection task, to determine the best and safest course of action. For example, moving vehicles in order to make space for fall protection equipment, especially in confined and/or areas with high density of other vehicles and equipment, adds additional hazards that may outweigh those of the fall hazard. In this case, personnel shall use deliberate risk management and other administrative controls to mitigate fall hazards. If the Competent Person for fall protection determines that the use of fall protection equipment is not feasible or practicable based on the overall hazard analysis, personnel shall use deliberate risk management and other administrative controls to mitigate fall hazards in accordance with guidance of the Competent Person for fall protection. A primary objective of the Fall Protection and Prevention Plan, is to provide site specific, standard methods to mitigate fall hazards associated with routine tasks and to provide guidance, should deviation to standard methods be required. The following paragraphs list types of fall protection systems that can be used for vehicle maintenance and inspection, in cases where fall hazards cannot be eliminated. Units must determine those fall protection measures which must be taken, based on risk analysis of the work or inspection to be performed:

Mobile Work Platforms (MWP). Where work is performed from elevated work platforms four feet or higher, the work platforms shall be equipped with a standard guardrail and safe method of access, or other fall protection system that mitigates potential fall hazards. Mobile servicing platforms are authorized, but shall be required to provide additional fall protection equipment per the manufacturer's specifications. MWP's provide powered access from self-propelled elevating platforms which are approved as Ground Support Equipment.

Other Aerial Work Platforms.
- B-Stands
- Phase Stands
- Warehouse Stands
- Ladders

Restraint System. A system consisting of equipment and components connected together designed to restrain a person from reaching an exposed fall hazard.

Personal Fall-Arrest System. A system used to arrest a person during a fall from a working level. It consists of an anchorage system, connecting means, and full-body harness and may include a lanyard, self-retracting device, lifeline, or suitable combination of these. A personal fall arrest system shall be rigged so that Soldier/employee arrested fall will not come in contact with a lower level or object. Safety belts (body belts) shall not be used. Suitable anchor points for personal fall arrest systems include horizontal lifeline systems, moveable and/or fixed anchorages attached to a rigid rail or beam, and vacuum type anchors. Vacuum anchors shall not be used on composite or thin shelled vehicles. They shall only be used on thick shelled vehicles.

Administrative Controls. These are controls that reduce risks through specific administrative actions. Methods for implementing administrative controls include:
- Fall Hazard Awareness Training for personnel performing elevated work.
- Providing suitable warnings, markings, placards, signs and notices.
- Establishing written policies, programs, instructions, and SOPs.
- Conducting a deliberate RM with consideration of the following as a minimum.
  - Maintenance or inspection work duration
  - Environmental factors such as wind, rain, snow, or ice
  - Probability of fall
  - Hazards of fall (obstacles, height, etc.)
  - Slippery materials or substances on vehicle surfaces
  - Actual requirement for work to be completed
- Limiting the exposure to a hazard (by reducing the number of personnel, and/or the length of time personnel are exposed).

Inspection, Care, Maintenance and Storage of Fall Protection Equipment. The Fall Protection Leaders Guide provides guidance, checklists, and specific requirements for this topic. The fall protection program instruction must address specific inspection, maintenance, storage, and care procedures for the fall protection equipment possessed and/or utilized by the command. Manufacturers' instructions and recommendations may provide a good starting point for these requirements. Checklists, maintenance requirement cards, and pre-operational inspection cards should be developed for each piece of fall protection equipment by a Competent Person for fall protection or FPPM and approved by the responsible authority.

Rescue Procedures. The Fall Protection Leaders Guide provides guidance and templates for this requirement. A site-specific rescue plan shall be prepared in writing and maintained for all instances where personnel work at heights while utilizing harness-type fall protection systems. If the rescue will be performed by the fire department or other governmental jurisdictional agency, a written pre-incident plan is required. The rescue plan shall contain detailed procedures on the methods of rescue to include methods of self-rescue/assisted rescue, equipment used in the rescue procedure, training requirements including specialized training for rescuers, procedures for requesting rescue, and a pre-mishap medical plan should medical assistance be required. The rescue plan should be included as part of the written Fall Protection and Prevention Plan.

Audits and Evaluations. The Fall Protection Leaders Guide provides compliance checklist for the Fall Protection Program which can be used for auditing the program. Program audits should be conducted semiannually, but shall be conducted annually as part of the command safety self-assessment.

Examples of Fall Protection Program Document Templates. The following are templates to assist units to establish, implement and manage viable fall protection programs. These templates include the following:
- Written Fall Protection Program Template
- Fall Hazard Survey Report Template
- Fall Protection and Prevention Plan Template

Chapter 14 – TIE-OFF CONSIDERATIONS AND SELECTION OF SAFE ANCHORAGES

One of the most important aspects of personal fall-arrest is fully planning the system before it is put in use. Probably the most overlooked component of the fall-arrest system is planning for suitable anchorages. Such planning should ideally be done during the design stage and before a structure or a building is constructed so that anchorages can be incorporated and identified during construction for maximum use later for maintenance work. If needed, properly planned and designed anchorages used during construction work may alsobe used afterward during maintenance provided they are installed and properly located for performing the maintenance task.

The strength of a personal fall-arrest system depends on its subsystems and components, as well as the anchorages and how strongly such a system is attached to the anchorage. Such attachment shall not significantly reduce the strength of the system, including the structural members (e.g., the beams or columns to which it is attached). If a method of attachment is used that will reduce the strength of the system, such component (e.g., beam or column) shall be replaced with a stronger one in order to maintain the appropriate maximum characteristics in compliance with IBC and design criteria documents.

- Lanyards shall not be connected to themselves or other lanyards unless permitted by the manufacturer.
- Knots shall not be tied in lanyards, lifelines, or anchorage connectors (i.e., anchor straps). Tie-off using a knot in a lanyard, lifelines, or anchorage connectors can reduce the strength by 50% or more.
- Tying a rope lanyard or lifeline around rough or sharp edges such as beams, columns or other surfaces may reduce the strength of the line due to cutting action of the sharp edge. If a line is cut or damaged, it will drastically affect the design reaction of the system during a fall. Such tie-off should be avoided or alternate rigging method should be used. As an alternate, use beam clamp, wire rope, effective padding, or abrasion resistance strap (chaffing protection) around or over the sharp or rough surfaces.
- The anchorage location should be as high as possible to minimize the free-fall distance and prevent any contact with an obstruction or the ground below if a worker falls. Free-fall distance shall not exceed six feet unless a specially designed lanyard is used that will allow the 12 foot free-fall provided the maximum

arresting force does not exceed 1,800 pounds. The anchorage point height shall reflect this restriction.

- Tie-off point(s) shall be located in such a way to minimize the swinging of the worker (pendulum-like motion) that can occur during a fall. The farther away in a horizontal direction a worker moves from a fixed anchorage (tie-off point), the greater the swing angle if a fall occurs. If any obstruction exists in the path of the swing fall, the force generated can be significant. The maximum angle of swing away from the tie-off point should not be more than 15 degrees in either direction.
- The strength of an eyebolt is rated along the axis of the bolt and its strength is greatly reduced if the force is applied at an angle to this axis (out-of- the-plane of the eye). Also, the diameter of the eyebolt should be compatible to snap-hook or carabiner attachment. Non-rotating rings should be avoided, since falls rarely occur directly along the axis of the eyebolt. Where possible, rotating rings (swivel rings) with full motion in the three axes should be used to increase the angle with the axis to more than 45 degrees. The ring will then be able to automatically align along the direction of force. Swivel rings used as anchorages in a fall arrest system shall be properly sized. The eyebolt used in the fall protection system shall be forged steel. Effort shall be made to minimize the angle between the axis of the eyebolt and the direction of the pull.
- Attaching two snap-hooks to the same anchorage: Where two Soldiers/employees are planning to use the same anchorage simultaneously by using two snap-hooks, the anchorage must be certified and rated for use by two people. Connecting both snap-hooks to the anchorage will require the use of additional connector.
- Horizontal lifelines, depending on their geometry and angle of sag, may be subjected to greater loads than the impact load imposed by an attached component. When the angle of sag for the horizontal lifeline is less than 30 degrees, the impact force generated is greatly amplified. For example, with a sag angle of 15 degrees, the force amplification is about 2:1 and at 5 degrees sag, it is about 6:1. Depending on the angle of sag, and the line's elasticity, the strength of the horizontal lifeline and the anchorages to which it is attached should be increased a number of times over that of the lanyard. Extreme care should be taken in considering a horizontal lifeline for multiple tie-off. The reason for this is that in a multiple tie-off to a horizontal lifeline, if one Soldier/employee falls, the movement of the falling worker may cause other Soldiers/employees to alsofall. Horizontal lifeline and anchorage strength should be calculated for each additional Soldier/employee to be tied-off. For these and other reasons, horizontal lifelines shall only be designed, selected, and certified by Qualified Person for fall protections. Inspection of installed horizontal lifelines and anchors before use is recommended.
- Considerations when evaluating horizontal lifeline systems.
  - Review the design calculations of the system.
  - Review manufacturer's test data of similar systems.
  - The anchorage and anchorage connector shall be compatible.
  - When tying off to a beam or column, do not attach the anchorage connection to a hole in the beam unless evaluated by a Qualified Person for fall protection, because the forces generated by a fall will weaken the beam structure. Do not drill a hole for tying off. This attachment will weaken the beam. The most favorable way to tie-off is to use an anchorage connection to wrap around the beam or column, such as an anchor strap, or use a designed beam clamp.
  - Do not tie a knot in the anchorage connection.
  - The most favorable location to tie-off to a beam is in the center of the span. This action will distribute the forces evenly at the supports. The closer the tie-off point is to the beam support, the shear-force of a fall on the structure will increase accordingly.
  - However, when tying-off to the beam always consider the hazard of swing fall effect.
  - Take into consideration the impact of shear forces at the supports and the bending moment distribution of forces beyond the supports into other structural members.
  - In the selection of a point of anchor in a column, take into consideration the effect of all forces due to axial loading and bending stresses.
  - Refrain from welding the anchorage connection to the anchorage, unless the welding is performed and certificated annually by a certified welder.
  - Where nails are used to install roof anchors, the number, type, and size of nails used to attach the component to a wood structure shall be in accordance with the building code requirements. Verify that the roof anchors are attached to structural members, rather than decking only.
  - Always specify the number of end-users that are allowed to attach to a specific anchorage.
  - In selection of anchorage location, take into

consideration the accessibility and ease of
securing or attaching to it (ease of tying off).
 - When attaching of the fall-arrest system to a concrete
slab, make sure the concrete is strong and thick
enough to sustain the static and dynamic loads
of the fall forces. The bottom steel reinforcement
in the concrete slab is usually under tension.
Concrete alone is very weak under tension.

Fall-Arrest System Considerations. Prior to
selecting a fall arrest system, the following
information should be verified and considered:
 - Movement of the worker
 - Existing obstructions in the worker's path
 - Location and availability of safe anchorages
 - Total fall distance, free fall distance
   and available clearance
 - Possibility of swing fall hazards
 - Compatibility of all components of the system
 - Impact forces
 - Availability of rescue/self-rescue

Chapter 15 –
GUIDANCE FOR FALL RESCUE PROCEDURES

Introduction. A person working at heights, using Fall
Protection equipment, may require rescue if that person
falls and is suspended in a harness. Prompt rescue is very
important. Studies show that a person suspended in a
harness may have blood circulation problems within a few
minutes. Accordingly, a site-specific Rescue Plan must
be prepared in writing and maintained for all instances
where personnel working at heights are exposed to fall-
hazards. An example plan is contained in Appendix F. The
Rescue Plan shall contain detailed procedures on the
methods of rescue; methods of self-rescue; equipment
used; training requirements; specialized training for
rescuer(s); procedures for requesting rescue; and available
medical assistance. Where the rescue may not be, or
cannot be, solely performed by a jurisdictional public
(e.g., city fire department) and/or government-emergency
response agency (e.g., government fire department), then
the Rescue Plan must contain detailed procedures for
planned rescue methods. The Rescue Plan is a part of the
written Fall Protection and Prevention Plan and contains

provisions for potential self-rescue or assisted rescue of
an end-user of fall protection. The Fall Protection and
Prevention Plan covers every fall-hazard to which end-users
are exposed. An example of a rescue plan is contained
in Appendix F - Site-Specific Fall Arrest Rescue Plan.

Background. Following a fall from a height, the end-user of
fall protection who is wearing a full-body harness properly
secured to an anchorage may be suspended in the harness
for a length of time if self-rescue or assisted rescue by co-
workers cannot be performed quickly. Sustained immobility
in a full-body harness may lead to suspension trauma
alsoknown as harness induced pathology. Suspension
trauma resulting from the accumulation of blood in the
veins is commonly called venous pooling. The symptoms
(known as orthostatic intolerance) of suspension trauma
include light-headedness, dizziness, weakness, and
occasionally, fainting. Normally, when an individual faints
and collapses onto a flat surface, the pooled blood now
no longer is being held down by gravity, and returns to
the heart, where it is once again distributed to the body.
Assuming no injuries are caused during the collapse,
the individual will quickly regain consciousness, and
recovery is likely to be rapid. When an individual hangs
in a harness in a vertical or near-vertical position without
leg motion, the same thing can happen; however, in this
case when consciousness is lost, the person remains
vertically suspended. An accumulation of blood in the
legs reduces the amount of blood in circulation. After an
initial speeding up of the heart-beat, the heart rate then
slows, and blood pressure will diminish in the arteries. The
reduction in quantity and/or quality (oxygen content) of
blood flowing to the brain leads to unconsciousness and
harmful effects on other vital organs. If these conditions
continue, they potentially may be fatal. The importance
of a timely rescue of a worker suspended in a harness, or
who has become incapacitated due to an injury and/or
heart attack, mandates the need for a written rescue plan.

General Requirements. Before an end-user of fall
protection is exposed to a fall-hazard, and before starting
work activities, the Competent Person for fall protection
and the end-user shall ensure there are pre-incident
plans and rescue plans in place that address the rescue

of a person who has fallen and becomes suspended in a harness. A checklist to determine if a person is competent or qualified is included in Appendix G. If a pre-incident plan is not available, the Competent Person for fall protection may work with the base safety office/officer, to obtain information from the jurisdictional public/government-emergency response agency, including emergency contact phone numbers and rescue capability. They then shall include this information in the rescue plan, along with alternative/supplemental rescue methods required to perform a timely rescue of an end-user suspended in a body harness, or one who is incapacitated at heights for other reasons. End-users of fall protection shall be trained in the methods for minimizing the effect or delaying suspension trauma if an end-user is suspended in a body harness and unable to perform a self-rescue and needs to wait to be rescued (e.g., keep legs moving and raise knees to the body, to help prevent the pooling of blood in the legs). Suspension straps attached to the harness can be used to minimize the effect of suspension trauma while the user is waiting for rescue. A strap for each leg is recommended. All end-users should be trained in the safe use of the straps.

Initiation of Rescue. End-user using fall protection equipment shall have an assigned safety person (spotter), also known as the "buddy system", who is within visual and aural range of the end-user. The assigned safety person must check periodically (at least every 5 minutes) to assure the end-user has not fallen and become suspended in the harness. The assigned safety person shall have the capability to make quick contact with the emergency response agency; or the end-user (or the team leader of a group of end-users) shall have this capability, in the case of the end-user or team visiting another Army activity.

Fall-Arrest Rescue Plan and Procedures. A site-specific rescue plan (for a Soldier/employee suspended in a full-body harness after a fall) shall be prepared in writing by the Competent Person for fall protection. In the case of the end-user or team visiting another Army activity, the rescue plan shall include the following:

- Pre-incident Planning. Per the NFPA 101 Life Safety Code, a written pre-incident plan is prepared by the jurisdictional public (e.g., city fire department) and/or governmentoemergency response agency (e.g., government fire department). Pre-incident planning is ensuring that responding emergency personnel knows as much as they can about a facility's construction, occupancy, and fire protection systems before an incident occurs. With this knowledge, the fire department can compare a potential incident at the facility with its available resources and plan the department's response accordingly. Pre-incident planning is not restricted to building components. It includes other factors and conditions that may be relevant to an emergency at a particular site. The end-user (or team leader of a group of end-users) in consultation and coordination with the Competent Person for fall protection shall verify that rescue procedures are in place for any workplace where the authorized rescuer will perform a rescue. The types of fall protection systems being used and the work environment shall be reviewed with the emergency response agency. The pre-incident plan shall be reviewed and updated by the activity's Competent Person for fall protection annually, or whenever there is a change to the job-site that will affect items in the plan.

- Methods of Rescue.
  - Jurisdictional Public Emergency-Response Agency.
  - Government Emergency-Response Agency.
  - Assisted Rescue: The written rescue plan shall include instructions for contacting rescue personnel, plus a description and verified location of all equipment to be used by the rescue team (e.g., scissor lift or aerial lift), and complete instructions and procedures for performing rescue safely and promptly.
  - Self-rescue. An end-user who has fallen and is suspended in a full-body harness and not incapacitated can usually perform a self-rescue, where the following conditions exist:
    - The end-user can reach an adjoining structure, and has the strength and mobility to pull up and onto the structure.
    - The end-user has a self-deploying or manual-deploying coiled webbing rescue ladder attached to lanyard anchorage, which after a fall allows climbing up to the anchorage point (or at least simply standing on the ladder, allowing the necessary circulation of blood to the entire body while an assisted rescue is being commenced).
    - An automatic or manual controlled descent device can be used as a self-rescue device if it is attached to a separate anchorage point (minimum 3,000-pounds strength) and a vertical tag-line is attached to the controlled descent device's safety snap-hook which

can be reached by the Soldier/employee suspended in the full-body harness. The tag-line is pulled, bringing down the self-retracting lanyard from the controlled descent device, and the descent device safety snap is attached either to the back D-ring or front rescue D-ring of the full-body harness, and the deployed shock absorber lanyard detached (this method is only viable if there is a "quick release" device which will allow the disconnecting of the shock absorber lanyard under tension). Once the deployed energy absorbing lanyard is disconnected from the full-body harness, the controlled descent device will allow the end-user to descend at a controlled rate to a lower level. This method requires "hands-on" training.

- Rescue Inspection. The inspection of equipment used by emergency response agencies is the responsibility of those agencies. Prior to use, the end-user shall inspect the self-rescue and assisted-rescue equipment to ensure that it is in safe working condition, and has been protected against damage from the weather (e.g., UV, water) and from workplace conditions (e.g., chemical, physical). Annually, a Competent Person for fall protection shall verify that the rescue equipment markings and instructions are consistent with ANSI and OSHA Standards, and that the rescue equipment has been maintained in accordance with manufacturer's instructions.

- Training Requirement for Rescue. Training is required for self-rescue techniques. All personnel who will work at height utilizing fall protection equipment shall be trained in self-rescue techniques. They shall be trained in these techniques before utilizing fall protection equipment and every two years thereafter.

- Specialized Training for the Rescuers. Training of rescue personnel at emergency response agencies are the responsibility of those agencies. For assisted-rescue, the authorized rescuers shall be properly trained and shall be proficient at performing a rescue of a person suspended in a harness or who has become incapacitated at heights. The authorized rescuer shall be knowledgeable in the selection, use, storage, and care of all equipment necessary to perform rescue on end-users from all types of fall protection equipment. Carefully evaluate hazards associated with rescue and determine whether or not it is safe to perform rescue. The authorized rescuer shall conduct a site visit to the work location prior to writing a post-fall-arrest rescue plan. The authorized rescuer shall assign and delineate various responsibilities in the rescue and evacuation of a Soldier/employee who has become incapacitated at heights and/or who is suspended in a body harness after a fall. Authorized rescuer training shall be conducted once every two years and evaluated at least annually by a Competent rescuer and shall include the following:
  - Fall-Hazard recognition, elimination and control methods.
  - Applicable fall protection and rescue regulations and standards.
  - Understanding and using the Fall Protection and Prevention Plan, and the Rescue Plan.
  - Inspection and maintenance of the equipment including manufacturers' instructions.
  - Proper uses of various rescue equipment.
  - Practical applications and drilling scenarios for rescue (Hands-on Training).

- Procedures for Requirement Requesting Rescue and Medical Assistance. The telephone number for emergency response agencies in CONUS is usually 911, or 9-911, depending upon the location. If the emergency response number is different, it must be posted and publicized throughout the worksite.

- Transportation Routes to a Medical Facility. A sketch indicating the route to the nearest medical facility/hospital should be included in the fall-arrest rescue plan and should be posted at the job site.

- Anchors Used for Rescue.
  - Anchorages selected for rescue systems including control descent devices shall be capable of sustaining static loads applied in the direction permitted by the rescue system of at least 3,000 pounds when designed as a rescue system only. If the anchorage for fall-arrest system is selected as a

rescue anchorage, it shall be capable of sustaining
5 times the foreseeable loads (certified rescue
anchorages), applied in the directions permitted by
the personnel fall-arrest system per attached person.
- Anchorage connectors used for rescue shall not be
attached to anchorages where such attachment would
reduce the allowable capacity of the anchorage itself.
- Anchorage connections shall be stabilized to
prevent unwanted movement or disengagement
of the rescue systems from the anchorage.
The rescue system shall be load tested before
a live load is placed on the system.
- The anchorage should be located at a point
above the rescuer to prevent swing fall.

• Fall-arrest Rescue Plans. The Fall-arrest Rescue
Plan should include the following information as
part of the Fall Protection and Prevention Plan:
- Provide a detailed location of the worksite, with any
information that will help find the location, building
number, floor number; etc. Post written directions
that can be read over the telephone to an ambulance
driver/police/fire department, or their dispatchers, on
how to get to the site from the main gate of a facility.
Give complete, accurate information to the rescue
responder. Post a map at the job-site, and highlight
with a yellow marker, the route one should take
from the site to the nearest hospital where someone
with minor injuries can be treated expeditiously.
- Indicate the location of the lift or other
equipment that will be used in case of
emergency, and the location of the key.
- Provide a detailed location of the closest first

aid kit. To assure that no time be lost looking for
first aid kits during an emergency, post a site
map marking the location of the first aid kits.
- List emergency telephone numbers. If an emergency
rescue is required, call the telephone numbers in
the order listed; i.e., 1st, 2nd, and 3rd. Post written
directions that can be read over the telephone
to an ambulance driver/police/fire department
or their dispatchers on how to get to the site
from the main gate of the facility. Give complete,
accurate information to the rescue responder.
- Send an escort to meet the fire department
upon arrival at the scene, and help them or the
rescuer find the location of the accident.
- Give the name of the person (the escort designated
to meet the fire department upon arrival at
the scene) and the back-up person (in case the
designated person is injured) responsible to
make the phone call in case of emergency.
- Indicate names of personnel that may require
rescue during the course of performing their jobs.
- If self-rescue is used, indicate the type of self-
rescue equipment available at the job site, or
which will be utilized during rescue operations.
- Indicate the training the rescuer should receive
in order to become a qualified rescuer.
- Initiate a buddy system when personnel are
working at heights and may require rescue. If the
buddy system is not feasible, contact the activity
to set up a visual or aural contact with the person
exposed to fall-hazards every five minutes.

Appendix A

SAMPLE FALL PROTECTION PLAN FOR BUILDING CONSTRUCTION
FOR

This Fall Protection Plan is specific to the following project:

Job Location:	
Date Plan Prepared:	
Date Plan Modified:	
Plan Prepared by:	
Plan Approved by:	
Plan Supervised by:	

2. STATEMENT OF UNIT POLICY

Unit Name is dedicated to the protection of its Soldiers/ employees from occupational injuries. All Soldiers/ employees of Unit Name have the responsibility to work safely on the job. The purpose of this plan is to supplement our existing Fall Protection Program and to ensure that every Soldier/employee who works for Unit Name recognizes workplace fall hazards and takes the appropriate measures to address those hazards.

This Fall Protection Plan addresses the use of conventional fall protection at a number of areas on the project, and identifies specific activities that require non-conventional means of fall protection. During the construction of wood frame (residential) buildings under 48 feet in height, it is sometimes infeasible or creates a greater hazard to use conventional fall protection systems at specific areas or for specific tasks. Such areas or tasks include, but are not limited to:

- Setting and bracing of roof trusses and rafters;
- Installation of floor sheathing and joists;
- Roof sheathing operations; and
- Erecting exterior walls.

In these cases, conventional fall protection systems may not be the safest choice for this project. This plan is designed to enable Soldiers/employees to recognize fall hazards associated with this job and to establish safe procedures to prevent falls to lower levels through holes and openings in walking/working surfaces.

Employer Responsibilities
- Ensure that all Soldiers/employees understand and adhere to the procedures of this plan and the instructions of the crew supervisor or foreman.
- Assign a Competent Person to be responsible for managing this Fall Protection Plan.
- Provide appropriate fall protection to Soldiers/ employees as detailed in this plan.

Soldier/Employee Responsibilities
- Bring to the attention of Unit Name management any unsafe or hazardous conditions or practices that may cause injury to themselves or other Soldiers/employees.
- Report any incident which causes injury to self or a co-worker.
- Each Soldier/employee will be trained in these procedures and will be expected to strictly adhere to them except when doing so would expose him/her to a greater hazard. If, in the Soldier's/employee's opinion, the procedures in this plan pose a risk, the Soldier/ employee is to notify Responsible Person and have their concern(s) addressed before proceeding with work.

Plan Manager Responsibilities. Responsible Person **shall function as manager of this Fall Protection Plan and has the following responsibilities:**
- Implement this fall protection plan.
- Perform continual observational checks of work operations to identify hazards.
- Enforce the unit policy and the procedures of this plan.
- Coordinate with crew supervisors or foremen to correct any unsafe practices or conditions immediately.
- Provide training on this plan to all affected Soldiers/ employees before work begins on this project.

Fall Protection to be Used on This Job. Installation of roof trusses/rafters, exterior wall erection, roof sheathing, floor sheathing, and joint/truss activities will be conducted by Soldiers/employees who are specifically trained to do this type of work and are trained to recognize fall hazards. The nature of such work normally exposes Soldiers/ employees to fall hazards for a short period of time. This Plan details how Unit Name will minimize these hazards.

Controlled Access Zones. When using this plan to implement the fall protection options available, workers must be protected through limited access to high hazard locations. Before any non-conventional fall protection systems are used as part of this work plan, a controlled access zone (CAZ) shall be clearly defined by a Responsible Person as an area where a recognized hazard exists. The demarcation of the CAZ shall be communicated by Responsible Person in a recognized manner, either through signs, wires, tapes, ropes, or chains.

- Unit Name shall take the following steps to ensure that the CAZ is clearly marked or controlled by a Competent Person.

- All access to the CAZ shall be restricted to authorized entrants only.
- All workers who are permitted in the CAZ must be listed in the appropriate sections of this plan (or be visibly identifiable by Responsible Person prior to implementation).
- Responsible Person shall ensure that all protective elements of the CAZ be implemented prior to the beginning of work.

Installation of Roof Truss or Rafter Erection

- During the erection and bracing of roof trusses/ rafters, conventional fall protection may present a greater hazard to workers. On this job, safety nets will not provide adequate fall protection because the nets will cause the walls to collapse. In addition, there are also no suitable attachment or anchorage points for guardrails or personal fall arrest systems.
- Requiring Soldiers/employees on this job to use a ladder for the entire installation process will cause greater hazard because the worker must stand on the ladder with his back or side to the front of the ladder. While erecting the truss or rafter, the worker will need both hands to maneuver the truss and therefore cannot hold onto the ladder. In addition, ladders cannot be adequately protected from movement while trusses are being maneuvered into place. Soldiers/employees may experience fatigue because of the increased overhead work with heavy materials, which can also lead to a greater hazard.
- Exterior scaffolds cannot be utilized on this job because the ground, after recent backfilling, cannot support the scaffolding. In most cases, the erection and dismantling of the scaffold would expose workers to a greater fall hazard than the erection of the trusses/rafters.
- On all walls 8 feet or less in height, Soldiers/employees will install interior scaffolds along interior walls below the location where the trusses/rafters will be erected. A sawhorse scaffold constructed of 46-inch sawhorses and 2 foot by 10 foot planks will often allow workers to be elevated high enough to allow for the erection of trusses and rafters without working on the top plate of the wall.
- In structures that have walls higher than 8 feet and where the use of scaffolds and ladders would create a greater hazard, safe working procedures will be used when working on the top plate, which will be monitored by Responsible Person. During all stages of truss/rafter erection, the stability of the trusses/rafters will be ensured at all times.
- Unit Name shall take the following steps to protect workers who are exposed to fall hazards while working from the top plate installing trusses/rafters:
  - Only trained and approved workers will be allowed to work on the top plate during roof truss or rafter installation. A list of approved Soldiers/employees will be maintained by the Responsible Person as an attachment to this plan.
  - Soldiers/employees shall have no other duties to perform during truss/rafter erection procedures.
  - All trusses/rafters will be adequately braced before any worker will be permitted to use the truss/rafter as a support.
  - Soldiers/employees will remain on the top plate using the previously stabilized truss/rafter as a support while other trusses/rafters are being erected.
  - Soldiers/employees will leave the area of the secured trusses only when it is necessary to secure another truss/rafter.
  - The first two trusses/rafters will be set from ladders leaning on sidewalls at points where the walls can support the weight of the ladder.
  - A Soldier/employee will climb onto the interior top plate via a ladder to secure the peaks of the first two trusses/ rafters being set.
  - Soldiers/employees responsible for detaching trusses from cranes and/or securing trusses at the peaks traditionally are positioned at the peak of the trusses/ rafters. There are also situations where workers securing rafters to ridge beams will be positioned at the top of the ridge beam. Unit Name will take the following steps to protect workers who are exposed to fall hazards while securing trusses/rafters at the peak of the trusses/ridge beam:
    - Only trained and approved workers will be allowed to work at the peak during roof truss or rafter installation. A list of approved Soldiers/employees will be maintained by Responsible Person as an attachment to this Plan.
    - Once truss or rafter installation begins, workers not involved in that activity shall not stand or walk below or adjacent to the roof opening or exterior walls in any area where they could be struck by falling objects.
    - Soldiers/employees shall have no duties other than securing/bracing the trusses/ridge beams.
    - Soldiers/employees positioned at the peaks, in the webs of trusses, or on top of the ridge beam shall work

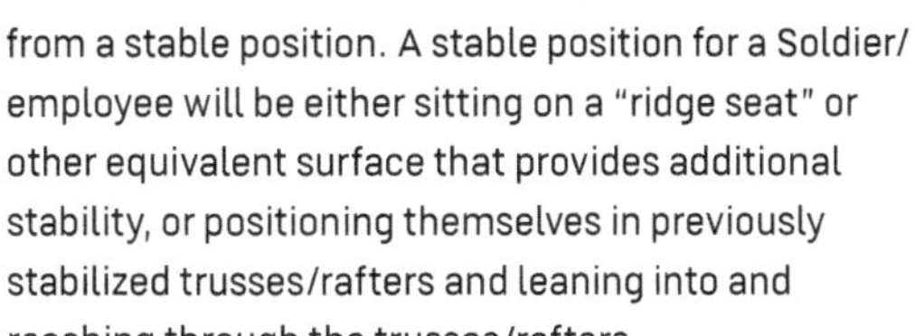

from a stable position. A stable position for a Soldier/employee will be either sitting on a "ridge seat" or other equivalent surface that provides additional stability, or positioning themselves in previously stabilized trusses/rafters and leaning into and reaching through the trusses/rafters.

- ◆ Workers shall not remain on or in the peak/ridge any longer than necessary to safely complete the task.

Roof Sheathing Operations

- Workers typically install roof sheathing after all trusses/rafters and any permanent truss bracing is in place. Because roof structures are unstable until some sheathing is installed, workers installing roof sheathing cannot be protected from fall hazards by conventional fall protection systems until it is determined that the roofing system can be used as an anchorage point. At that point, Soldiers/employees shall be protected by personal fall arrest systems.
- Trusses/rafters are subject to collapse if a worker falls while attached to a single truss with a belt/harness. Nets could alsocause collapse, and there is insufficient structure to attach guardrails.
- All Soldiers/employees will ensure that they have secure footing before they attempt to walk on the sheathing, and will clean their shoes/boots of mud or other slip hazards.
- To minimize the time workers must be exposed to a fall hazard, materials will be staged to allow for the quickest installation of sheathing.
- Unit Name will take the following steps to protect workers who are exposed to fall hazards while installing roof sheathing:
  - Once roof sheathing installation begins, Soldiers/employees not involved in that activity shall not stand or walk below or adjacent to the roof opening or exterior walls in any area where they could be struck by falling objects.
  - Responsible Person shall determine the limits of this area, which shall be clearly communicated to workers prior to placement of the first piece of roof sheathing.
  - Responsible Person may suspend work on the roof for brief periods as necessary to allow other workers to pass through such areas when this would not create a greater hazard.

- Only trained and approved workers will be allowed to install roof sheathing. A list of approved Soldiers/employees will be maintained by Responsible Person as an attachment to this plan.
- The bottom row of roof sheathing may be installed by workers standing in truss webs.
- After the bottom row of roof sheathing is installed, a slide guard extending the width of the roof shall be securely attached to the roof. Slide guards will be at least 4 inches in height and capable of limiting the uncontrolled slide of workers. Workers shall install the slide guard while standing in truss webs and leaning over the sheathing.
- Additional rows of sheathing may be installed by workers positioned on previously installed rows of sheathing with slide guards.
- Additional slide guards shall be securely attached to the roof at intervals not to exceed 13 feet as successive rows of sheathing are installed. For roofs with pitches in excess of 9 to 12, slide guards will be installed at 4 foot intervals.
- When wet weather conditions (rain, snow, or sleet) are present, roof-sheathing operations shall be suspended unless safe footing can be assured for those workers installing sheathing.
- When strong winds (over 40 miles per hour) are present, roof-sheathing operations shall be suspended unless windbreakers are erected.

Installation of Floor Joists and Sheathing. Unit Name will take the following steps to protect workers who are exposed to fall hazards while installing floor joists or floor sheathing:

- Only trained and approved workers will be allowed to install floor joists and floor sheathing. A list of approved Soldiers/employees will be maintained by Responsible Person as an attachment to this plan.
- Materials for this work shall be conveniently staged to allow for easy access to workers.
- The first-floor joints or trusses will be rolled into position and secured either from the ground, ladders, or sawhorse scaffolds.
- Each successive floor joist or trust will be rolled into place and secured from a platform created from a sheet of plywood laid over the previously secured floor joists or trusses.

- Except for the first row of sheathing, which will be installed from ladders or the ground, Soldiers/employees shall work from the established deck.
- Any Soldiers/employees not assisting in the leading edge construction while leading edges still exist (i.e., cutting the decking for installers) shall not be permitted within 6 feet of the leading edge under construction.

Erection of Exterior Walls. Unit Name will take the following steps to protect workers who are exposed to fall hazards during the construction and erection of exterior walls:

- Only trained and approved workers will be allowed to construct and erect exterior walls. A list of approved Soldiers/employees will be maintained by Responsible Person as an attachment to this plan.
- A painted line 6 feet from the perimeter will be clearly marked prior to any wall erection activities to warn of the approaching unprotected edge.
- Materials for operations shall be conveniently staged to minimize fall hazards.
- Workers constructing exterior walls shall complete as much cutting of materials and other preparation as possible away from the edge of the deck.

Enforcement. Constant awareness of and respect for fall hazards, as well as compliance with all safety rules, are considered conditions of employment with Unit Name.

The crew supervisor or foreman, as well as Responsible Person or unit management, reserve the right to issue disciplinary warnings to Soldiers/employees, up to and including termination, for failure to follow the guidelines of this plan.

Mishap Investigations. All incidents that result in injury to workers and near misses, regardless of their nature, shall be reported and investigated IAW AR 385-10. All incidents shall be investigated as soon as possible by Responsible Person to identify the cause and means of prevention to prevent future occurrences. In the event of such an incident, this Fall Protection Plan shall be reviewed to determine if additional practices, procedures, or training should be implemented to prevent similar incidents in the future.

Changes to the Plan. Any changes to this plan will be made by Responsible Person. This plan shall be reviewed by Responsible Person as the job progresses to determine if additional practices, procedures, or training are needed to improve or provide additional fall protection. Affected Soldiers/employees shall be notified of changes to this plan, and retrained, if necessary. A copy of this plan and all approved changes shall be maintained at the jobsite by Responsible Person.

Appendix B
SAMPLE FALL HAZARD SURVEY REPORT

General information

Facility Activity:________________________________
____________________________________Page #___
Building/Facility #______________________________
____________________________________Date: ___
Department: ___________________________________
Work Area: ____________________________________
Survey Conducted by: ___________________________
Accompanied by: _______________________________

Survey Data
Fall Hazard Zone and Type:______________________
____

Work Location: _________________________________
____
Personnel Interviewed: __________________________
____
Applicable Regulations/Standards:_________________
____
Type of work performed: _________________________
____
How close is the person to the fall hazard: __________
____
Location and distance to obstructions:______________
____

Suggested anchorage location, if fall hazard cannot be eliminated or prevented: ___________________

____

Available clearance and total fall distance: _____________

Number of personnel exposed to fall hazard: ____________

Frequency and duration of exposure: __________________

Exposure rating: High _______ Medium ________ Low _______

Potential severity of a fall: ___________________________

____

Any obstructions in the potential fall path:______________

____

Access or egress to fall hazard area: __________________

____

Condition of floor or other surfaces: _________________

____

Review any mishap reports at the facility:______________

____

Any chance of slips trips and same level falls:
Yes _____No _____

Lock-Out/Tag-Out hazard: ____________________________

____

Floor/Surface Condition: _____________________________

____

Identify the presence of:
❑ Hot objects:______________________________________
❑ Sparks: ____
❑ Flames: ____
❑ Heat producing objects:____________________________
❑ Any electrical/chemical hazards: ___________________
❑ Sharp objects: ___________________________________
❑ Abrasive surfaces: _______________________________

❑ Any moving equipment in the area: ___________________
❑ Impact of weather factors: _________________________
❑ Other maintenance work environment/issues: __________

Suggested fall protection solutions:
Select two of the following probable solutions
❑ Guardrails ______________________________________
❑ Safety nets _____________________________________
❑ Fall arrest system_______________________________
❑ Travel restraint system __________________________
❑ Work positioning system _________________________
❑ Horizontal lifeline system/vertical lifeline ____________
❑ Aerial lift equipment/work platforms ________________
❑ Warning line system _____________________________
❑ Ladder climbing devices __________________________
❑ Raising/lowering devices _________________________
❑ Covers ____

If fall arrest/restraint/work positioning/HLL system is selected:
❑ Anchorage(s) location (if any): _____________________
❑ Can rescue be performed if required: ________________
❑ Type of rescue: _________________________________
❑ Any potential swing fall hazards: ___________________
❑ Is the end-user properly trained: Yes ____ No ____
❑ Other factors: __________________________________

Any additional information: __________________________

____

Drawings/Sketches/Photos

Prepared by: _____________________________________

Approved by: _____________________________________

FALL PROTECTION PROGRAM COMPLIANCE CHECKLIST FOR PERSONNEL PERFORMING WORK AT HEIGHTS, EXPOSED TO FALL HAZARDS AND USING FP EQUIPMENT	Date of Audit:
Project	Name
Prepared/Audited by	Signature

FALL PROTECTION PROGRAM CRITERIA	Yes	No	N/A
1 Does the project have personnel working at heights, exposed to fall hazards above 4 or 6 feet and using fall protection (FP) equipment? Is there a possibility of a fall from any height onto dangerous equipment, into a hazardous environment or onto an impalement hazard? Is there any need to deviate from the 4 or 6-foot threshold requirement? Is this deviation approved by the designated Competent Person? If **Yes**, a Fall Protection Program is required to be established and implemented.			
BASIC PROGRAM REQUIREMENTS			
2 Is the Fall Protection Program written and approved by the project?			
3 As an alternate to the written Fall Protection Program, is the project using the Fall Protection Guide as their program with Safety Office approval?			
ADDITIONAL REQUIREMENTS			
4 Is there a need for the project to have additional requirements above and beyond the requirements stated in this Guide?			
DUTIES AND RESPONSIBILITIES			
5 Did the project delineate duties and assigned responsibilities of personnel involved in the Fall Protection Program, including Program Manager, Competent and Qualified Persons for fall protection, in the implementation of a managed Fall Protection Program?			
6 Do the assigned personnel have the necessary skills, knowledge, training and expertise to manage, administer, and implement the Fall Protection Program safely?			

WORKPLACE SURVEYS AND ASSESSMENT OF FALL HAZARD			
7 Has a survey been conducted for each fall hazard at existing buildings, facilities or structures and a Fall Hazard Survey Report prepared?			
8 Was fall hazard analysis performed to determine the risk assessment, hazard severity and fall mishap probability?			
9 Are one or more fall protection methods identified in the survey report to eliminate or control each fall hazard?			
10 Do the surveyed walking working surfaces have the structural integrity to safely support the workers (i.e. working on roofs)?			
11 For personnel conducting roof inspections and, investigations, have they received proper training to conduct the work safely, prior to accessing the roof?			
FALL PROTECTION AND PREVENTION PLAN			
12 For personnel exposed to fall hazards and using fall arrest equipment (not otherwise protected by passive fall protection system such as guardrails) has a Site Specific Fall Protection and Prevention Plan been prepared and submitted to the Safety Office for review and approval? (It is recommended to prepare a generic fall protection and prevention plan for non-routine tasks (i.e. emergency tasks)). The plan shall be updated as conditions change, once every six months.			
13 Is the Fall Protection and Prevention Plan prepared either by the designated Competent or Qualified Person for fall protection? If the plan includes fall protection components or systems requiring direction, supervision, design calculations or drawings by the qualified person for fall protection, the name, qualifications and responsibilities of the qualified person shall be addressed in the plan.			
14 Does the plan describe in detail the specific practices, equipment, methods and procedures to be used for the protection of workers from falling to lower level and the inspection requirements?			
PREFERRED ORDER OF CONTROL MEASURES			
15 Have the fall hazards been evaluated to determine the preferred order of control measure for selecting the appropriate fall protection method (i.e. elimination or prevention)?			
16 Can fall hazards be eliminated by alternate work methods or changing task(s) or process(s)?			
SELECTION OF FALL PROTECTION MEASURE			
17 Is the most appropriate fall protection method selected, compatible with the type of work being performed?			

STANDARD GUARDRAIL SYSTEM				
18	If guardrails are used, do they comply with the specified requirements for height, strength and minimum material of construction?			
19	If perimeter cables used at unprotected side or edge, as a method of attaching a lanyard to the cables, do they meet the design requirements for horizontal lifelines? Did the qualified person for fall protection design the system as a horizontal lifeline system?			
COVERS				
20	If covers are used to cover a hole 2 inches in its least dimension, are they capable of withstanding without failure, at least twice the combined weight of the worker, equipment and material? When covers are used, are they clearly marked or color coded?			
WORK PLATFORMS				
21	When working from elevated work platform, is the platform equipped with guardrail or other fall protection system? Is the work platform maintained properly?			
SAFETY NET SYSTEM				
22	Does the safety net installation meet the specified criteria and requirements, including the size of the mesh openings and the strength of the outer rope or webbing			
23	Has the safety net been tested in a suspended position immediately after installation and under the supervision of a qualified person?			
24	If a safety net is relocated, repaired or left in place for more than 6 months, was it retested in suspension under the supervision of a qualified person?			
25	Is the inspection of the safety net performed by a Competent Person and in accordance with manufacturer's recommendations?			
26	Inspection of safety nets shall be performed immediately after installation, weekly thereafter, and following any alteration or repair. Has the inspection been documented?			
PERSONAL FALL PROTECTION SYSTEMS				
27	Do all the fall arrest systems and equipment used meet ANSI/ASSE Z359 Fall Protection Code Standards?			
28	When selecting personal fall protection system, are the free fall distance, total fall distance and available clearance taken into consideration?			
29	Do the snap-hooks and carabiners used meet ANSI Z359 Fall Protection Code/Standards? (Snap-hooks and carabiners meeting ANSI Z359.1-1992 (R1999) shall not be used.)			
30	For workers having body weight outside the capacity range of 130-310 lbs. and using fall protection equipment, is it permitted in writing by the manufacturer?			

31	If it is necessary to increase the free fall distances beyond 6 feet (i.e. tying at the foot level) and limiting the maximum arresting force on the body under 1,800 lbs., is the qualified person for fall protection making this determination? When the tie off point is located below the dorsal D-ring, use the 12-foot free fall energy absorbing single or "y" lanyards. A qualified person for fall protection is required to make this determination.			
32	If the sternal D-ring attachment point of the body harness (located at the sternum) is used for fall arrest, is the worker exposed to a free fall distance of less than two feet and the average arrest force on the body is 900 lbs.?			
33	Self-retracting devices (SRD) shall not be used in horizontal applications unless permitted by the manufacturer. Is the SRD used in vertical application?			
34	When using "y" lanyard for 100% tie-off, does the joint between the two legs of the lanyard withstand a force of 5,000 lbs.?			
35	The unused leg of the "y" lanyard shall not be attached to any part of the harness, except to attachment points specifically designated by the manufacturer. Has the manufacturer of the equipment designated such attachment points?			
36	When using positioning system, is the worker using a separate system (secondary system) that provides back-up protection from a fall? When using a restraint system, is the lanyard length short enough (or adjustable) to prevent a worker from being exposed to a fall hazard?			
37	When using ladder climbing devices for ascending or descending on fixed ladders, is the length of connection point between the body harness and the rail or cable 9 inches or less? Will the system be activated within two feet after a fall occurs? Prior to installation, has the ladder to which the climbing device will attached to, been designed to withstand the forces generated by the fall of the climber?			
	FALL ARREST EQUIPMENT SELECTION CRITERIA			
39	Does the selected fall arrest equipment meet ANSI Z359 (2007) Fall Protection Code? **(Any equipment meeting ANSI A10.14 shall not be used)**			
40	Can the manufacturer of the selected equipment substantiate thru Third-Party Testing Laboratories, Witness Testing or Manufacturer Self-Certification Testing that the equipment meets ANSI Z359 Fall Protection Code/Standards and/or designed, selected and approved by the qualified person for fall protection?			
	TRAINING			
41	Is all fall protection training for all personnel involved in the fall protection program in accordance with the ANSI Z359 Fall Protection Code?			

#				
42	Are workers trained by a Competent Person for fall protection who is qualified to deliver the training on the safe use of fall protection and rescue equipment, including hand s on and practical demonstrations?			
43	Did the assigned Competent and Qualified Persons for fall protection receive adequate training?			
44	Did other personnel involved in the fall protection program receive adequate training?			
45	Has the above training been documented and verified with a certificate of training?			
46	Did end-users receive refresher/update training on the use of fall protection equipment once every two years? Did the Competent Person for fall protection receive refresher/ update training to stay current with the fall protection and educational requirements once every two years? Did other personnel involved in the fall protection program receive recommended or required refresher/update training as specified in ANSI Z359.2 Standard?			
	SELECTION OF ANCHORAGES FOR FALL ARREST EQUIPMENT			
47	For fall arrest anchorages selected/identified and designed by a qualified person for fall protection, are they capable of supporting at least twice the maximum arresting force? For fall arrest anchorages selected by a Competent Person for fall protection, are they capable of supporting a minimum force of 5,000 pounds per person attached?			
48	For positioning and travel restraint anchorages selected by a Competent Person for fall protection are they capable of supporting 3,000 pounds per Soldier/employee attached? If positioning and restraint anchorages selected and designed by a qualified person for fall protection, do they meet the requirement of two times the foreseeable force on the worker?			
49	Are the horizontal lifeline anchorages designed by a registered professional engineer with experience in designing HLL systems; or designed by a qualified person for fall protection who has appropriate training and experience?			
	RESCUE PLAN AND PROCEDURES			
50	For personnel working at heights and using fall arrest equipment, has a site-specific rescue plan and procedures been prepared and maintained at the work location?			
51	If self-rescue or assisted-rescue are the planned methods to be used during rescue, did personnel conducting rescue receive adequate training?			
52	If required, are independent anchorages for rescue identified and selected?			
53	If the method of rescue is by the jurisdictional public and government-emergency response agencies, has a pre-incident plan been developed?			

INSPECTION OF PERSONAL FALL PROTECTION EQUIPMENT			
54	Have procedures been established for inspection, storage care and maintenance of the equipment and IAW manufacturer's instructions and recommendations?			
55	Does the Competent Person for fall protection inspect the fall protection equipment annually and w/documentation?			
56	Does the end-user inspect the equipment prior to each use?			
FALLS FROM HEIGHTS MISHAP REPORTING			
57	Are falls from heights mishaps reported to the Safety Office?			
EVALUATION OF PROGRAM EFFECTIVENESS			
58	Are procedures in place to audit and evaluate the fall protection program, at least once every two years?			

Appendix D
FALL PROTECTION EQUIPMENT INSPECTION CHECKLIST

Project:	
Inspected by:_________________________ (Competent Person's Name)	Date:
Work Area:	Other:

Instructions:
1. All parts of the fall protection system and components are to be checked for excessive wear and damage.
2. Use the symbol "Y" for yes or OK.
3. Use the symbol "N" for no or replace.
4. All equipment must be inspected visually before each use by the end-user and by the Competent Person at least annually with documentation.

Name or Equip #	Self-Retracting Lifelines		Lanyards		Full-Body Harnesses			Horizontal Lifeline System		
	Cable	Mechanism	Webbing	Energy Absorber	Webbing	D-Rings and Connectors	Labeling	Anchorage Connection/ Stanchions	Cable	Hardware

Prepared either by the Competent Person or a person trained and designated by the Competent Person for fall protection.

Anchorages
• Do workers know appropriate anchorage points for each task that requires a fall-arrest/positioning or restraint system? ____

• Are all anchorage points stable, substantial, and have sufficient strength to withstand twice the potential impact energy of the free-fall? ___________________________

• Is the D-ring of the full-body harnesses located at the back shoulder height? ___________________________

• Are anchorage points for self-retracting lifeline systems located overhead? ___________________________

• Can the Soldier/employee move from one station to another or climb up and down without exposure to a fall? ____

• If the lifeline, lanyard, or self-retracting lifeline is not permanently attached to an anchorage point at the elevated work area, is the first worker up or the last worker down protected while climbing and traversing?___________

Lanyards
• Is the lanyard length as short as necessary and in no case greater than 6 feet? (1.8 meters) ___________________

• Are manually adjustable lanyards used when it is desirable to be able to take slack out of the lanyard? ____

• Does the lanyard have a shock-absorbing feature to limit the arresting forces? ___________________________

• If the lanyard has a shock absorber, is it obvious to the user that the shock absorber has been deployed? (Is there a warning label, broken pouch, etc.) ___________________

• Have you prohibited tying of knots from the lanyard to the lifeline? (Mechanical rope grabs or fall arresters must be used) ____

Self-Refracting Lanyard (SRL)
• Are Soldiers/employees properly trained to use a SRL? ____

• Is the SRL under a regular maintenance and inspection program?____

• Is the end of the cable properly spliced? ______________ (Thimble eye, flemish eye-spliced, and swaged fitting/ferrule)

Snap-Hooks
• Are double-locking snap-hooks being used? ___________

• Is the snap-hook attached to the D-ring, eyebolt, or other hardware in a manner approved by the manufacturer of the snap-hook? ___________________________________

• Are snap-hooks inspected regularly for stress, wear, distortion, and spring failure? ___________________

• Are snap-hooks arranged so they are never connected to each other? ___________ (They should NOT be connected to each other).

Full-Body Harnesses
• Are full-body harnesses selected for a particular job equipped with all necessary attachment points? (For fall arresting, work positioning, descent control, rescue, or ladder fall-protection systems)___________________

• Are body harnesses inspected regularly for wear, abrasion, broken stitching, and missing hardware? ____

• Is the velcro type of closure prohibited from all load-bearing connections?___________________________

• Have workers been instructed in the use and care of body harnesses? ____

Fall Arresters
• Is the fall arrester compatible with the lifeline on which it is to be installed or operated? _______________________

• Is the fall arrester in operational condition? ___________

• Is the fall arrester equipped with a changeover lever that allows it to become a stationary anchor on the lifeline?
____

• Is the fall arrester equipped with a locking mechanism that prevents unintentional opening of the device and subsequent disengagement from the lifeline? ___________

• Is the fall arrester's "up" direction marked properly so that the equipment can be attached to the line correctly?
____

Vertical Lifelines
• Does the lifeline have a minimum breaking strength of 5,000 pounds? (2,268 kilograms) _____________________

• Is the lifeline protected from abrasive or cutting edges?
____

• Does the system provide fall protection as the worker connects to and releases from the lifeline? _____________

• Is the lifeline arranged so workers never have to hold it for balance? (A lifeline should never be used for balance)
____

• Is the vertical segment integrated with the horizontal segment to provide continuous fall protection?___________

Horizontal Lifelines
• Has the entire horizontal lifeline system been designed and approved by a qualified person? ________________

• Have the anchorages to which the lifeline is attached been designed and evaluated specifically for a horizontal lifeline? ____
• Has the designer of the system approved the number of Soldiers/employees that will be using it? ________________

• Is the rope or cable free from signs of wear or abrasion?
____

• Does the rope or cable have the required initial sag?
____

• Have the workers been warned about potential falls? _____
Have the clearances been checked? ___________________

• Is the hardware riding on the horizontal lifeline made of steel? (Aluminum is not permitted because it wears excessively) _____________________________________

• Is the fall arrester included in a regular maintenance and inspection program? ______________________________

Other Considerations
• Has the free-fall distance been considered, so that a worker will not strike a lower surface or object before the fall is arrested? ___________________________________

• Have pendulum-swing fall hazards been eliminated?
____

• Have safe methods to retrieve fallen workers been planned? _______________________________________

• Is all fall-arrest equipment free of potential damage from welding, chemical corrosion, or sandblasts? _____________

• Are all components of the system compatible according to the manufacturer's instruction? ____________________

• Have Soldiers/employees been properly trained in the following issues? ___________________________________

• Manufacturer's recommendations, restrictions, instructions, and warnings _____________________________

• Location of appropriate anchorage points and attachment techniques ____

• Are there any problems associated with elongation, deceleration distance, and method of use, inspection, and storage? ____

• Are all regular inspections performed by trained inspectors? _____________________________________

• Are written reports maintained?_____________________

• Has the total fall distance been considered?___________

• Has rescue of the worker been considered? ___________

Appendix F
SITE-SPECIFIC FALL ARREST RESCUE PLAN CHECKLIST

GENERAL INFORMATION				
Project:				Date:
Building/Facility #:			Primary and Secondary Phone Numbers:	
Detailed Location:			Ladder/Lift Location:	
First Aid Kit Location(s):	1		Fire Extinguisher Locations:	1
	2			2
	3			3
Nearest Medical Facility and Directions:				
Procedure for Requesting Rescue:	1			
	2			
	3			
	4			
Describe Rescue Operation and Method:				
Types of Equipment used (Ladder, Hoist, Aerial Lift, etc.):				
If Self-Rescue or Assisted Rescue is Planned, Describe Equipment to be used:				
Specialized Training for the Rescue Team:				

Describe if Additional Anchorages for Rescue are Required:	
Has Rescue Plan been developed in coordination with Local Emergency Services (essential if relying on them to provide rescue)?	
Is a pre-incident plan prepared when the planned Method of Rescue is the Fire Department?	
Additional Comments:	
Prepared By:	
Approved By:	

Optional Checklist for use when a Competent or Qualified Person is required to develop fall protection and prevention plan

COMPETENT OR QUALIFIED PERSON CHECKLIST				
Project:	Location:			
FP Program Manager:	Date:			
COMPETENT PERSON INFORMATION				
Competent Persons Name: ____ Length of experience in this occupation: _________________ Length of experience with this employer: _______________				
TRAINING KNOWLEDGE AND EXPERIENCE		Yes	No	N/A
Does the designated individual have training knowledge and experience in:				
Applicable fall protection regulations, standards and requirements?				
Fall hazard recognition (How to recognize and identify fall hazards?).				
Duties and responsibilities of other designated personnel under the FP Program (e.g. qualified person, end-user, authorized rescuer, etc.)?				
Conducting fall hazard surveys and preparing survey report?				

The requirements and criteria for guardrails, safety nets, scaffolds, aerial lifts and movable and stationary work platforms, warning line system, and safety monitoring system?			
Developing fall protection and prevention plans (written fall protection procedures)?			
Notes:			

1. If the Fall Protection and Prevention plan includes fall protection components or systems requiring direction, supervision, design calculations or drawings by a Qualified Person for Fall Protection or a professional engineer, the name, qualifications, responsibilities, training knowledge, experience and signature of the Qualified Person for Fall Protection or professional engineer shall also be addressed in the plan. 2. At a minimum, the qualified person/professional engineer information is required when using horizontal lifelines, other engineered systems, the anchorages or tie off points are located below the dorsal D–ring and designing certified anchorages that require being twice the maximum arrest or potential force.			
Fall arrest, positioning, restraint and ladder climbing systems.			
Fall hazard elimination and control methods including how to assemble, disassemble and use fall protection systems and equipment (donning of the equipment, equipment installation techniques and proper anchoring and tie-off techniques)?			
Fall protection system and equipment assessments (e.g. component compatibility, estimating free fall distances, total fall distance and required clearance, and common hazards of each system and component used and determining when a system is unsafe.			
How to conduct detailed inspection storage care and maintenance of equipment, components and systems with documentation?			
Fall protection rescue equipment and procedures and prepare fall hazard rescue plan?			
The selection and use of non-certified anchors (e.g. 5,000 lbs. anchorage for fall arrest)?			
Requirements for working over or near water or working from/in machinery over water.			
List training/experience including certificate of training;			

AUTHORITY	Yes	No	N/A
Does the designated individual have authority from the project to:			
Take prompt corrective action to eliminate existing and predictable hazards?			
Stop work?			

All personnel signing this form indicate they understand the fall hazards on the job site, and that they have been trained in the proper use of and will use the selected fall protection equipment and methods. Review and sign again if hazards, methods or work changes.

Course Title		
Course Description		
Instructor Name		Date
Instructor Signature		Duration of Training
Student Information		
Name		Signature

Unit Name INSTRUCTION XXXX.XX

From: Commanding Officer, Unit Name

Subj: FALL PROTECTION PROGRAM MEMORANDUM

Encl: (1) Fall Hazard Survey Report
(2) Fall Protection and Prevention Plan

Purpose. The purpose of this instruction is to establish a UNIT Fall Protection Program, provide policy and requirements for the implementation of the program, and establish procedures on fall protection and fall prevention for Unit Name personnel working at heights and/or exposed to fall hazards while conducting aircraft maintenance and inspection work.

Applicability. This program applies to all Unit Name military and civilian personnel who are exposed to fall hazards when performing maintenance or inspection work on an elevated, walking, or working surface with unprotected sides, edges, or openings, from which there is a possibility of falling four feet or more to a lower level; or where there is a possibility of a fall from any height onto dangerous equipment, into a hazardous environment, or onto an impalement hazard.

Cancellation. Instruction XXXX.XX

Background. Falls from heights are a major cause of injuries and fatalities in the workplace. The nature of aviation maintenance and inspection requires that Unit Name personnel work at heights, thereby exposing them to fall hazards.

Unit Fall Protection Policy. Unit Name is committed to providing a safe working environment for its personnel exposed to fall hazards and eliminating preventable mishaps. Mission accomplishment is our number one task, but we cannot accomplish this task if we Do not do our utmost to ensure the safety of our personnel. This program is part of an overall unit safety program designed to enhance operational readiness by preventing injury or death of personnel through careful management of material resources.

- Unit Name personnel shall take every reasonable precaution to protect themselves and others while working at heights.
- Units may use the Fall Protection Program Guide, as a guide when creating their own site-specific program, plans, and policies.

Duties and Responsibilities. Unit Name leadership shall ensure that all personnel assigned to the fall protection program have the necessary skills, knowledge, training, and expertise to manage, administer, and implement the fall protection program.

- Unit Safety Officer: Shall provide oversight of the Unit Fall Protection Program.
- The Unit Fall Protection Program Manager shall:
  - Develop and implement the unit fall protection program.
  - Manage and coordinate the unit's core Fall Protection program.
  - Perform and document reviews, and evaluations of operations, facilities, materials, and equipment.
  - Conduct fall hazard survey and prepare a survey report.
  - Coordinate workplace and fall-arrest system inspections with the supporting Competent Person for Fall Protection.
    - The Competent Person support services can be requested from the supporting regional/local installation safety office.
    - The Competent Person services may be provided by unit personnel if they have completed the required training course.
  - Correct or abate identified fall hazards.
  - Obtain consulting services from the supporting regional/local installation safety office on technical aspects of the program.
  - Coordinate with the supporting regional/local installation safety office to analyze core effectiveness through the annual safety self-assessment process.
  - Coordinate core fall protection training for all personnel who are exposed to fall hazards and end-users of fall-arrest systems, when applicable.
  - Coordinate with the supporting regional/local installation safety office any other support required to ensure the proper management of this program.

Fall-hazard Prevention and Control. Units are required to survey the workplace to identify potential fall hazards and prepare a "Fall Hazard Survey Report". A "Fall Protection and Prevention Plan", prepared by a Competent Person for Fall Protection, or by a Qualified Person for Fall Protection, is required as part of a managed fall protection program. The trained FPPM may also prepare the fall protection and prevention plan

- The Fall Protection and Prevention Plan shall provide site-specific guidance for tasks executed at Unit Name.
- Prior to visiting a site at another Army Activity, Unit Name Soldiers/employees who will be climbing or using ascending equipment different from the equipment addressed in the Unit Name Fall Hazard Survey Report should review that the host Activity's Fall Hazard Survey.

Fall Protection Training. Unit Name personnel who have the potential for exposure to fall hazards and/or those involved in the Fall Protection Program shall be trained in fall prevention and fall protection. A Competent Person or a vendor who has the knowledge, expertise, and education to deliver the training shall train end-users.

- Initial Training. Unit personnel exposed to fall hazards shall complete the following training:
  - Fall hazard awareness training: general fall hazard awareness training shall be provided by a Competent Person for Fall Protection.
  - Personnel required to climb aircraft shall be qualified climbers.
    - Qualified climbers shall be trained and knowledgeable regarding those surfaces which are, and are not, able to support a climber; and shall observe airframe restrictions. Training shall include:
    - Requirement for head protection
    - Areas on airframe authorized for use as a step
    - Checking to ensure that steps are cleaned of slippery substances
    - Instruction to maintain three points of control on aircraft at all times
    - Qualified climbers shall rely on experience, proper training, and the use of time critical risk management to minimize the risk of a fall.
  - End-users of fall protection shall also under go "Hands-On" training on the specific personal fall-protection system and equipment used at the unit. This training must be performed by a Competent Person for Fall Protection. End-user training shall include training on:
    - Safe use of equipment
    - Proper application of equipment
    - Equipment limits
    - Estimation of fall distances and clearance requirements

- Methods of inspection, storage, care, and maintenance
  - Applicable regulations
  - Proper anchoring and tie-off techniques
  - Recognition of fall-hazard deficiencies
    w Site-specific procedures.
- Retraining. Retraining in relevant topics shall be provided to the end-user and/or qualified climbers when:
  - Personnel have been observed using maintenance stands or fall protection equipment in an unsafe manner.
  - Personnel have been involved in a mishap or a near-miss incident.
  - Personnel have received an evaluation that reveals he or she is not using the fall protection equipment properly.
  - An end-user is assigned a different type of fall protection equipment.
  - A condition in the workplace changes in a manner that could affect the safe use of the fall protection equipment that the end-user is to utilize.
- Refresher training. Personnel exposed to fall hazards shall receive periodic awareness and hands-on equipment (when applicable) refresher training.

Inspection, Storage, Care, and Maintenance of Fall Protection Equipment. Unit fall protection equipment shall comply with the manufacturer's requirements for the inspection, storage, care and maintenance of fall protection equipment.

- A Competent Person for fall protection or a Competent Person for equipment inspection shall perform a documented inspection on each piece of fall protection equipment annually.
- End-users of fall protection shall inspect fall protection equipment prior to each use.

Mishap Reporting. Any fall-from-height experienced by unit personnel shall be reported to the FPPM or Safety Officer. If a fall results in fall-arrest equipment activation, the event shall be reported as a near miss using the ASMIS 2.0 system.

Program Audits and Evaluation. The FPPM shall audit the program annually (at a minimum).

Rescue Plan. When fall-arrest systems are utilized the FPPM shall coordinate with the supporting Competent Person for fall protection to create a rescue plan which meets the requirements in Appendix E of the Fall Protection Leaders Guide. The rescue plan shall then be incorporated into the unit Fall Protection Program.

(Signed)
UNIT C.O.

Appendix J
UH-60 FALL HAZARD SURVEY REPORT (Template)

Purpose. The purpose of this survey report is to provide a baseline assessment of fall protection and associated hazards on the UH-60 aircraft, and points for consideration when executing tasks required for aircraft maintenance and inspection. This analysis designates three distinct locations for aircraft maintenance, three distinct working areas on the aircraft itself, and designates tasks to be performed in these areas as high risk or low risk. Any fall, even a fall on the same level, can result in disability or death; but for the purpose of this analysis, fall severity shall be generalized, based on potential fall distance.

- Falls of 4 feet or less are considered relatively minor.
- Falls from 4-10 feet are considered moderate falls that likely would result in bone fractures, severe contusions, and a lost-time injury.
- Falls greater than 10 feet are considered severe and may result in severe injury including partial disability or death.
- Falls from any height where there is potential of falling onto hazardous conditions (e.g., machinery) are considered severe.

Mitigation Method. Fall protection and fall protection methods are parts of the overall hazard analysis and prevention plan for an aviation maintenance and/or inspection evolution. Location of the aircraft or potential fall exposure, nature of the task, work area of the aircraft or working platform, and consideration for other hazard potentials that may be introduced with the use of fall protection methods, shall be considered for each task. For example, moving aircraft in order to make space for fall protection equipment, especially in confined and/or areas with high density of other aircraft and equipment, may create additional hazards that outweigh those of the fall hazard.

- The use of fall protection systems and equipment on an active flight line may introduce foreign object damage (FOD) hazards and hazards associated with jet, propeller, and rotor wash.
- Consideration must be given to all hazards involved with the execution of a particular maintenance or inspection task to determine the best and safest course of action to protect personnel as well as equipment.

Hazard Analysis. Exposures to fall hazards are considered as the product of frequency of exposure, proximity to fall hazard, duration of exposure, security of area, and the environmental impact. These factors provide the overall probability of the fall.

- The hazard assessments provided in this document are displayed using tables which quantify fall hazards based on severity and probability of a hazard becoming a mishap.
- Where multiple fall hazards exist during a particular work task, the one with the highest associated risk (worst-case) shall be utilized to determine the proper fall hazard controls to use during the entire work task. A hazard assessment has been included for aircraft moves as a point for consideration in the overall hazard analysis of an aircraft maintenance or inspection.
- This assessment is not all-inclusive of all hazards or conditions that may arise, but is a guideline for supervisors and workers to execute operational risk management, to protect personnel and materiel.

Fall Hazard Areas.
- Tasks and RM hazard assessments
  - Tail Pylon
    - Fall Hazard: Height of total exposure is approximately 12 feet from the top of tail gearbox cowling to the ground. The top step is approximately 10 feet depending on aircraft weight and servicing of the tail strut.
    - High-Risk Tasks include:

- Removal, installation, and/or maintenance on the tail gearbox input seal and input flange.
- Removal, installation, and/or maintenance on the tail rotor paddles.
- Removal, installation, and/or maintenance on inboard and outboard retention plates.
- Removal, installation, and/or maintenance of No.6 driveshaft.
- Removal, installation, and/or maintenance on tail rotor quadrant.
- Low-Risk Tasks include:
- Preflight inspections to include daily and turnaround maintenance preflight inspections.
- Servicing of tail rotor gearbox.
- Resetting cam-locks on tail cowlings.
- Risk: Steps integrated into airframe provide access to tail and associated components. Steps may be inadequate to maintain balance and mitigate risk of falling while attempting to perform aircraft maintenance and inspection. The risk of falling increases significantly with less than three points of contact between the aircraft and the person performing aircraft maintenance or inspection.
- Hazard assessment:

HIGH-RISK TASK RAC MATRIX

Hazard Severity	Probability			
	A	B	C	D
I	1	1	2	3
II	1	2	3	4
III	2	3	4	5
IV	3	4	5	5

LOW-RISK TASK RAC MATRIX

Hazard Severity	Probability			
	A	B	C	D
I	1	1	2	3
II	1	2	3	4
III	2	3	4	5
IV	3	4	5	5

- Mitigations:
- Fall protection stand with appropriate PPE.
- Use of integrated aircraft steps with appropriate PPE.
- Deliberate RM/administrative controls. Some examples can include:
  - Consider specific part of aircraft used to climb onto aircraft and determine whether or not a safer alternate exists.
  - Check weather conditions and postpone aircraft climb if necessary.
  - Ensure aircraft surfaces are free of slippery substances.
  - Assure that boot soles are in good condition

- Assure the climbing personnel maintain a
  minimum of three points of contact at all times.
- Don head protection

Hydraulics Bay Area

- Fall Hazard: Height of total fall exposure is approximately
  seven feet from the top of the decking of the hydraulics
  bay to the ground. This height is dependent on
  aircraft weight and the servicing of the struts.
- High-Risk Tasks to be performed:
  - Torque verifications or adjustments on
    main rotor damper mount bolts.
  - Torque verifications or adjustments on
    bifilar assembly and weights.
  - Torque verifications or adjustments on
    the main gearbox mount feet.
  - Torque verifications or adjustments
    of the main rotor blades.
  - Removal and/or installation of forward bridge.
  - Removal and/or installation of main rotor blades.
  - Removal, installation, and/or
    maintenance on primary servos.
- Low-Risk Tasks include:
  - Preflight inspections, to include daily and
    turnaround maintenance preflight inspections.
  - Servicing of hydraulic reservoirs.
- Risk: Steps integrated into airframes provide access
  to the hydraulics bay, transmission, main rotor, and
  associated components. Steps and working surfaces
  may be inadequate to maintain balance and mitigate
  risk of falling while attempting to perform aircraft
  maintenance and inspection in these areas. The risk
  of falling increases significantly with less than three
  points of contact between the aircraft and the person
  performing aircraft maintenance or inspection.
- Hazard assessment:

HIGH-RISK TASK RAC MATRIX

Hazard Severity	Probability			
	A	B	C	D
I	1	1	2	3
II	1	2	3	4
III	2	3	4	5
IV	3	4	5	5

LOW-RISK TASK RAC MATRIX

Hazard Severity	Probability			
	A	B	C	D
I	1	1	2	3
II	1	2	3	4
III	2	3	4	5
IV	3	4	5	5

- Mitigations:
  - Fall protection stand with appropriate PPE.
  - Use of integrated aircraft steps with appropriate PPE.
  - Deliberate RM/administrative controls.
    Some examples can include:
    - Consider specific part of aircraft used to
      climb onto aircraft and determine whether
      or not a safer alternate exists.
    - Check weather conditions and during inclement
      weather postpone aircraft climb if necessary.
    - Ensure aircraft surfaces are free of slippery substances.
    - Assure boot soles are in good condition.
    - Assure that climbing personnel maintain
      minimum three points of contact at all times.
    - Don head protection.
- Area above Cabin
  - Fall Hazard: The height of area above the cabin
    is approximately nine feet from the top of the
    decking above the oil cooler compartment to
    the ground. This height is dependent on aircraft
    weight and the servicing of the struts.
  - High-Risk Tasks to be performed:
    - Removal and/or installation of the rescue hoist.
    - Torque verifications or adjustments on
      main rotor damper mount bolts.
    - Torque verifications or adjustments on
      bifilar assembly and weights.
    - Torque verifications or adjustments on
      the main gearbox mount feet.
    - Torque verifications or adjustments
      of the main rotor blades.
    - Removal and/or installation of aft bridge.
    - Removal and/or installation of main rotor blades.
    - Removal and/or installation of the APU.
    - Removal, installation, and/or
      maintenance on primary servos.
  - Low-Risk Tasks to be performed:
    - Preflight inspections to include daily and
      turnaround maintenance preflight inspections.
    - Servicing of utility hydraulic pump.
    - Servicing of the APU.
  - Risk: Steps integrated into airframe provide
    access to an area above the cabin to include the
    oil cooler compartment, main engines, main rotor,
    ECS compartment, fire bottle compartment, APU

compartment, and associated components. Steps may be inadequate to maintain balance and mitigate risk of falling while attempting to perform aircraft maintenance and inspection. The risk of falling increases significantly with less than three points of contact between the aircraft and the person performing aircraft maintenance or inspection.

- Hazard assessment:

HIGH-RISK TASK RAC MATRIX

Hazard Severity	Probability			
	A	B	C	D
I	1	1	2	3
II	1	2	3	4
III	2	3	4	5
IV	3	4	5	5

LOW-RISK TASK RAC MATRIX

Hazard Severity	Probability			
	A	B	C	D
I	1	1	2	3
II	1	2	3	4
III	2	3	4	5
IV	3	4	5	5

- Mitigations:
  - Fall protection work stand with appropriate PPE.
  - Use of integrated aircraft steps with appropriate PPE.
  - Deliberate RM/administrative controls. Some examples can include:
  - Consider the specific part of aircraft used to climb onto the aircraft, and determine if a safer alternate exists.
  - Check weather conditions, and during inclement weather postpone aircraft climb if necessary.
  - Ensure that aircraft surfaces are free of slippery substances.
  - Assure that boot soles are in good condition.
  - Assure that the person maintains a minimum of three points of contact at all times.
  - Don head protection.
- Additional Hazard analysis for consideration:
  - Aircraft move
    - Hazard: Aircraft Move - UH-60 aircraft weigh between 15,000 and 22,500 pounds. Moving aircraft can be accomplished with tow tractors or manually. Aircraft moves can be highly dynamic evolutions that put personnel at risk of being fatally run over or injured. Aircraft moves also have high incidences of causing aircraft damage which can be quite costly and adversely affect mission readiness.
    - Where aircraft are parked in close quarters, the hazards associated are increased. Close quarters also make the aircraft movement much more technically demanding for personnel steering the aircraft being moved.
    - Ground handling personnel may be unable to remain safely out of the path of the aircraft.
  - Hazard assessment:

Hazard Severity	Probability			
	A	B	C	D
I	1	1	2	3
II	1	2	3	4
III	2	3	4	5
IV	3	4	5	5

- Mitigations:
  - Use of appropriate, properly trained personnel.
  - Use of deliberate risk management prior to moving aircraft.
  - Movement of fall protection equipment.
  - Hazard: UH-60 aircraft have a main rotor diameter of over 53 feet 8 inches with blades that can droop to less than four feet under certain circumstances. Movement of fall protection stands could result in aircraft damage.
- Hazard assessment:

Hazard Severity	Probability			
	A	B	C	D
I	1	1	2	3
II	1	2	3	4
III	2	3	4	5

- Mitigations:
  - Use of appropriate, properly trained personnel to include spotters to mitigate contact between fall protection stands and rotor blades.
  - Use of deliberate RM prior to moving equipment.

The elements contained in this Fall Hazard Survey Report have been reviewed and approved.

______________(sign)_____________ _______________
Fall Protection Program Manager   Date
______________(sign) _____________ _______________
Competent Person for Fall Protection   Date

Description of fall hazards that may be encountered at the workplace by end-users during performance of their work. The UH-60 has three primary work areas that may expose personnel to fall hazards, as the working surfaces are greater than four feet above ground level. The working area forward of the main rotor mast shall be generalized and referred to as the hydraulics bay area. This area potentially exposes personnel to a fall hazard of seven feet. The working area aft of the main rotor head to the end of the fire bottle compartment shall be generalized as the oil cooler area and potentially exposes personnel to a nine-foot fall hazard. Though the tail rotor area is not a work platform per OSHA definition, work may be executed on the integrated steps, exposing personnel to a potential 10-foot fall hazard as measured from the top step.

Type of fall protection/fall prevention methods or systems used for every phase of work. Military aviation maintenance and inspection is inherently dynamic and fraught with potential hazards. OSHA requires fall protection when working four feet or more above lower levels to be mitigated by one or more of five control measures. Additionally, Federal regulation mandates this requirement, except where employers can demonstrate that such fall protection systems are infeasible or would create a greater hazard.

Fall protection control measures are parts of the overall hazard analysis and fall protection and prevention plan for aviation maintenance and/or inspection. Location of the aircraft or potential fall exposure, nature of the task, work area of the aircraft or working platform, and consideration for other hazard potentials that may be introduced with the use of fall protection control measures, shall be considered for each task.

- For example, moving aircraft in order to make space for fall protection equipment to execute a low risk task, especially in confined and/or areas with high density of other aircraft and equipment, adds additional hazards that may outweigh those of the unprotected fall hazard.
- Consideration must be given to all hazards involved with the execution of a particular maintenance or inspection task to determine the best and safest course of action.

- Fall hazards shall be mitigated according to the hierarchy of controls.

The overall approach to fall prevention at Unit Name shall be in the forms of prevention, engineering controls, and administrative procedures. As there are no anchor points or sufficient height above ground for proper employment of a fall-arrest system on the UH-60 airframe without the use of external equipment or an engineering change to the hangar, the use of fall-arrest system is not practicable at this time for Unit Name.

- Feasibility of a fall-arrest system for applications within the hangar shall be explored in the future at Unit Name.
- Unit Name has fall protection stands that provide both fall prevention, on the backside of stands with handrails, and engineering controls, on the aircraft side of the stands, to minimize the potential for falls. The squadron also has B-1 and B-4 work stands, phase stands, and ladders that provide some degree of fall prevention and engineering controls.
- There are some locations where the use of stands is not practicable, due to overall hazard analysis, or mission requirements. In these situations or areas, administrative controls shall be utilized. The primary administrative controls utilized at Command Name are the use of qualified climbers, and deliberate and time critical RM. Three primary aircraft locations shall be addressed in this plan: flight line, hangar bay, and wash rack.

Flight line. Fall protection equipment and systems generally are not authorized on active flight lines, due to the risk of foreign object damage (FOD), and the potential effects of rotor, jet, or propeller wash. Use of fall protection equipment in this environment may cause potential for greater hazards than the fall hazard itself.

- Fall protection equipment normally should not be utilized on an active flight line at Unit Name unless a risk management assessment indicates that the presence of the equipment does not create a larger overall hazard to personnel than the risk of a fall.

- Maintenance should be performed in a hangar whenever feasible.
- Aircraft inspection and maintenance on active flight lines shall require the use of deliberate and time critical risk management and restricting access to qualified climbers in order to minimize the risk of a fall.
- In the consideration of using maintenance stands and/or fall-arrest systems where fall potential might be increased by moving aircraft into a hangar, RM also should be applied.
- Consideration should be given to moving aircraft requiring high-risk maintenance actions into the hangar in order to utilize maintenance stands and/or fall-arrest systems to prevent a fall.

Wash rack: Maintenance stands with guards or arrest systems shall be used for all aircraft washing procedures where maintenance personnel are on the aircraft greater than four feet above ground level.

- Other maintenance and inspection tasks executed at the wash rack not associated with aircraft washing shall be executed using the same fall protection measures as normally would be used in the hangar.

Hangar Bay:
- Tail. Due to the nature of the tail steps and increased potential for falls from greater heights, fall protection maintenance stands (with guardrails at the entire periphery) shall be used to the maximum extent practicable on all maintenance actions to the tail where the potential for a fall of four feet or more exists. B-1, B-4, phase, and/or fall protection stands shall be configured and positioned properly to mitigate the fall hazard.
  - Should the use of a stand be deemed impracticable; a Competent Person for Fall Protection, and Maintenance Control shall approve deviations and determine alternative fall protection methods, case-by-case. Nonetheless, railed maintenance stands, arrest systems, or other fall protection measures should be used, except in rare exceptions.
  - Ladders alsomay be used for tasks determined low-risk by the fall hazard assessment, as long as they are used and set up properly. Personnel utilizing the ladder shall maintain three points of contact with the ladder at all times, and the ladder itself shall be positioned with four points of contact at all times.

Hydraulics bay area. The hydraulics bay area has a maximum fall exposure of seven feet. Risk assessment for the specific task to be executed can be found in the hazard assessment (see Fall Hazard Survey).

- Fall protection stands shall be used to the maximum extent practicable on all maintenance actions in the hydraulics bay area, where the potential for falling four feet or greater exists.
- Where B-1, B-4, phase, and/or fall protection stands are used, they shall be configured and positioned properly to mitigate the fall hazard, and to prevent the creation of additional fall hazards which may be caused by improperly configured stands.
- Should the use of a stand be deemed impracticable for a high-risk task as defined in the hazard assessment; a Competent Person for Fall Protection and a Maintenance Control representative shall approve deviations; and they shall determine alternative fall protection methods, case-by-case.
- Should the use of a stand be deemed infeasible for a low-risk task as defined in the hazard assessment (Fall Hazard Survey); the end-user, under the direction of a Competent Person for Fall Protection, may utilize administrative controls, which shall include deliberate risk management prior to execution of the task.

The use of administrative controls as the sole fall protection methodology is authorized only during the rare occasions when the movement of, setup, and use of a fall protection stand or arrest system creates a greater hazard than the hazards associated with the low-risk maintenance action to be performed. For example, the hazard of moving maintenance or fall-arrest system stands onto an active flight line may outweigh a low risk fall hazard. Normally, engineering controls (maintenance stands) or other controls to prevent a fall from occurring (fall-arrest system) always should be used.

Oil cooler area. The oil cooler area has a maximum fall exposure of nine feet. Risk assessment for the specific task to be executed can be found in the hazard assessment (Fall Hazard Survey).
- Work stands shall be used to the maximum extent practicable on all maintenance actions in the oil cooler area, where the potential for falling from a height of four feet or greater exists.
- If B-1, B-4, phase, and/or fall protection work stands are used, they shall be configured and positioned

properly to mitigate the fall hazard and prevent the creation of additional fall hazards which may be caused by improperly configured stands.

- Should the use of a stand be deemed impracticable for a high-risk task as defined in the hazard assessment; a Competent Person for Fall Protection, and a Maintenance Control member, must approve deviations and determine alternative fall protection methods, case-by-case.
- Should the use of a stand be deemed infeasible for a low-risk task as defined in the hazard assessment (Fall Hazard Survey) the end-user, under the direction of a Competent Person for Fall Protection, may utilize administrative controls which shall include deliberate RM at a minimum, and execute the task.
- The use of administrative controls as the sole fall protection method is authorized only during the rare occasions when the movement of, setup, and use of a fall protection stand or arrest system creates a greater hazard than the hazards associated with the low-risk maintenance action to be performed. For example, the hazard of moving maintenance- or fall-arrest system stands onto an active flight line may outweigh a low risk fall hazard.
- Normally, engineering controls (maintenance stands) or a fall-arrest system should be used.
- Training requirements for personnel exposed to fall hazards. Unit Name civilians and military personnel who use fall protection equipment (stands); and other Army personnel involved in the fall protection program; shall be trained in accordance with OSHA, in order to recognize, evaluate, and control fall hazards.
  - Initial Training. Command personnel exposed to fall hazards shall complete the following training:
    - Fall hazard awareness training: general fall hazard awareness training shall be provided by either a Competent Person for Fall Protection or by completion of an approved general fall hazard awareness training.
    - Personnel required to climb aircraft shall be qualified climbers.
    - Qualified climbers shall be trained and knowledgeable regarding what surfaces are and are not able to support a climber and shall observe airframe restrictions. Training shall include:
      - Requirement for head protection
      - Identification of areas on airframe authorized for use as a step
      - Checking to ensure that steps are cleaned of slippery substances
      - Methods to maintain three points of contact on aircraft at all times
      - Qualified climbers shall rely on experience, proper training, and the use of time-critical risk management to minimize their risks of a fall.
      - End-users of Fall-arrest Systems also shall undergo "Hands-On" training on the specific fall-arrest system equipment used at the command. This training must be performed by a Competent Person for Fall Protection. End-user training shall include training on:
      - Safe use of equipment
      - Proper application of equipment
      - Equipment limits
      - Estimation of fall distances and clearance
      - Proper anchoring and tie off techniques
      - Methods of inspection, storage, care, and maintenance
      - Applicable regulations
      - Recognition of fall-hazard deficiencies
      - Site specific procedures
  - Retraining. Retraining in relevant topics shall be provided to the end-user and/or qualified climbers where:
    - The end-user or qualified climber has been observed using fall protection equipment in an unsafe manner.
    - The end-user or qualified climber has been involved in a mishap or a near-miss incident.
    - The end-user or qualified climber has received an evaluation that reveals that he or she is not using the fall protection equipment properly.
    - The end-user or qualified climber is assigned a different type of fall protection equipment.
    - A condition in the workplace changes in a manner that could affect the safe use of the fall protection equipment that the end-user is to utilize.
  - Refresher training. Personnel exposed to fall hazards shall receive refresher training on the safe use of fall protection equipment at a minimum of every two years.
- Type of fall protection equipment and systems

provided to the Soldiers/employees that
might be exposed to fall hazards.
- Fall Protection Stands:
  ◆Aircraft Maintenance Platform Type B-1
  ◆Aircraft Maintenance Platform Type B-4
  ◆West Coast Weld Tech H-60 Helicopter
    Maintenance Platform
  ◆Metallic Ladder Mobile Work Platform
    SCS 8460 and SCS 9636-44
- Ladders: Ladders are not technically fall protection
  equipment though they may be used to ascend working
  or inspection areas at heights when the use of fall
  protection equipment is not feasible or not available.
- Fall-arrest Systems
  - When used as part of the fall protection program,
    fall-arrest systems (harnesses, lanyards,
    and associated hardware) shall be inspected
    annually or following the manufacturer's
    instructions whichever is more stringent.
  - Fall-arrest systems shall be inspected annually
    by a Competent Person for Fall Protection and
    inspection documentation shall be maintained
    by the FPPM for a period of at least five years.
- Current Competent Persons
  - Competent Person for Fall Protection: As
    designated by the commander and trained
    in accordance OSHA standards.
  - Competent Person for Equipment Inspection:
    As designated by the commander.
- Fall protection equipment and instructions for
  assembly, disassembly, storage, maintenance,
  and care. B-1 and B-4 maintenance railed
  platforms are considered support equipment
  and shall be assembled, disassembled, and
  maintained by the support equipment division.
  - Pre-operational inspections shall be conducted

prior to each use by the end-user in accordance
with manufacturer's instructions.
  ◆Phase and other maintenance stands not under the
    purview of the support equipment division shall
    be assembled, disassembled, and maintained per
    manufacturer instructions at a maximum one year
    interval by a Competent Person for Fall Protection
    or a Competent Person for Equipment Inspection.
  ◆Maintenance stands shall be inspected by
    the end-user using the Unit Name (local)
    pre-operational checklist. Fall protection
    maintenance stands shall be stored in
    accordance with manufacturers' instructions.
  - Ladders: Ladders shall be inspected by the end-user
    prior to each use, and inspected at maximum one-year
    intervals by a Competent Person for Fall Protection
    or a Competent Person for Equipment Inspection.
- Rescue plan and procedures. Unit Name does not
  currently employ the use of fall-arrest systems
  or equipment. Should this type of equipment be
  utilized in the future by Unit Name personnel, the
  FPPM shall coordinate Competent Person for Fall
  Protection support to create a rescue plan which
  meets the requirements in Appendix E of the Fall
  Protection Leaders Guide. The rescue plan shall then be
  incorporated into the command Fall Protection Program.
- The elements contained in this Fall Prevention and
  Protection Plan have been reviewed and approved.

```
_______________________________     _____________
Unit Fall Protection Program Manager      Date
_______________________________     _____________
Competent Person for Fall Protection      Date
```


Unit Name INSTRUCTION XXXX.XX

From: Commanding Officer, Unit Name

Subject: FALL PROTECTION PROGRAM MEMORANDUM

Encl: 1 Fall Hazard Survey Report
2 Fall Protection and Prevention Plan

Purpose. The purpose of this instruction is
to establish a Unit Fall Protection

Program. Provide policy and requirements for the
implementation of the program, and establish procedures
on fall protection and fall prevention for Unit Name
personnel working at heights and/or exposed to fall hazards
while conducting vehicle maintenance and inspection work.

Applicability. This program applies to all Unit Name military and civilian personnel who are exposed to fall hazards when performing maintenance or inspection work on an elevated, walking, or working surface with unprotected sides, edges, or openings, from which there is a possibility of falling four feet or more to a lower level; or where there is a possibility of a fall from any height onto dangerous equipment, into a hazardous environment, or onto an impalement hazard.

Cancellation. Instruction XXXX.XX

Background. Falls from heights are a major cause of injuries and fatalities in the workplace. The nature of vehicle maintenance and inspection requires that Unit Name personnel work at heights, thereby exposing them to fall hazards.

Unit Fall Protection Policy. Unit Name is committed to providing a safe working environment for its personnel exposed to fall hazards and eliminating preventable mishaps. Mission accomplishment is our number one task, but we cannot accomplish this task if we do not do our utmost to ensure the safety of our personnel. This program is part of an overall unit safety program designed to enhance operational readiness by preventing injury or death of personnel through careful management of material resources.

- Unit Name personnel shall take every reasonable precaution to protect themselves and others while working at heights.
- Units may use the Fall Protection Program Guide, as a guide when creating their own site-specific programs, plans, and policies.

Duties and Responsibilities. Unit Name leadership shall ensure that all personnel assigned to the fall protection program have the necessary skills, knowledge, training, and expertise to manage, administer, and implement the fall protection program.

- Unit Safety Officer: Shall provide oversight of the Unit Fall Protection Program.
- The Unit Fall Protection Program Manager shall:
  - Develop and implement the unit fall protection program.
  - Manage and coordinate the unit's core Fall Protection program.
  - Perform and document reviews, and evaluations of operations, facilities, materials, and equipment.
  - Conduct fall hazard survey and prepare survey report.
  - Coordinate workplace and fall-arrest system inspections with the supporting Competent Person for Fall Protection.
  - The Competent Person support services can be requested from the supporting regional/local installation safety office.
  - The Competent Person services may be provided by unit personnel if they have completed the required training course.
  - Correct or abate identified fall hazards.
  - Obtain consulting services from the supporting regional/local installation safety office on technical aspects of the program.
  - Coordinate with the supporting regional/local installation safety office to analyze core effectiveness through the annual safety self-assessment process.
  - Coordinate core fall protection training for all personnel who are exposed to fall hazards and end-users of fall-arrest systems, when applicable.
  - Coordinate with the supporting regional/local installation safety office any other support required to ensure the proper management of this program.

Fall-hazard Prevention and Control. Units are required to survey the workplace to identify potential fall hazards and prepare a "Fall Hazard Survey Report." A "Fall Protection and Prevention Plan," prepared by a Competent Person for Fall Protection, or by a Qualified Person for Fall Protection, is required as part of a managed fall protection program. The trained FPPM may alsoprepare the fall protection and prevention plan

- The Fall Protection and Prevention Plan shall provide site-specific guidance for tasks executed at Unit Name.
- Prior to visiting a site at another Army Activity, Unit Name Soldier/employees who will be climbing or using ascending equipment different from the equipment addressed in the Unit Name Fall Hazard Survey Report should review that the host Activity's "Fall Hazard Survey".

Fall Protection Training. Unit Name personnel who have the potential for exposure to fall hazards and/or those involved in the Fall Protection Program shall be trained in fall prevention and fall protection. A Competent Person or a vendor who has the knowledge, expertise, and education to deliver the training shall train end-users.

- Initial Training. Unit personnel exposed to fall hazards shall complete the following training:
  - Fall hazard awareness training: general fall hazard awareness training shall be provided by a Competent Person for Fall Protection.

  - Personnel required to climb vehicle shall be qualified climbers.
    - Qualified climbers shall be trained and knowledgeable regarding those surfaces which are, and are not, able to support a climber; and shall observe vehicle restrictions. Training shall include:
    - Requirement for head protection
    - Areas on vehicle authorized for use as a step
    - Checking to ensure that steps are cleaned of slippery substances
    - Instruction to maintain three points of control on vehicle at all times
    - Qualified climbers shall rely on experience, proper training, and the use of time-critical risk management to minimize the risk of a fall.
    - End-users of Fall Protection shall also undergo "Hands-On" training on the specific personal fall-protection system and equipment used at the unit. This training must be performed by a Competent Person for Fall Protection. End-user training shall include training on:
    - Safe use of equipment
    - Proper application of equipment
    - Equipment limits
    - Estimation of fall distances and clearance requirements
    - Methods of inspection, storage, care, and maintenance
    - Applicable regulations
    - Proper anchoring and tie-off techniques
    - Recognition of fall-hazard deficiencies
    - Site-specific procedures.

- Retraining. Retraining in relevant topics shall be provided to the end-user and/or qualified climbers when:
  - Personnel have been observed using maintenance stands or fall protection equipment in an unsafe manner.
  - Personnel have been involved in a mishap or a near-miss incident.

- Personnel have received an evaluation that reveals that he or she is not using the fall protection equipment properly.
- An End-user is assigned a different type of fall protection equipment.
- A condition in the workplace changes in a manner that could affect the safe use of the fall protection equipment that the end-user is to utilize.

- Refresher training. Personnel exposed to fall hazards shall receive periodic awareness and hands-on equipment (when applicable) refresher training.

Inspection, Storage, Care, and Maintenance of Fall Protection Equipment. Unit fall protection equipment shall comply with the manufacturer's requirements for the inspection, storage, care and maintenance of fall protection equipment.

- A Competent Person for Fall Protection or a Competent Person for Equipment Inspection shall perform a documented inspection on each piece of fall protection equipment annually.
- End-users of Fall Protection shall inspect fall protection equipment prior to each use.

Mishap Reporting.

- Any fall-from-height experienced by Unit personnel shall be reported to the FPPM or Safety Officer.
- If a fall results in fall-arrest equipment activation, the event shall be reported as a near miss using the ASMIS 2.0 system.

Program Audits and Evaluation. The FPPM shall audit the program annually (at a minimum).

Rescue Plan. When fall-arrest systems are utilized, the FPPM shall coordinate with the supporting Competent Person for Fall Protection to create a rescue plan which meets the requirements in Appendix E of the Fall Protection Leaders Guide. The rescue plan shall then be incorporated into the Unit Fall Protection Program.

(Signed)
UNIT C.O.

Purpose. The purpose of this survey report is to provide a baseline assessment of fall protection and associated hazards on the M978, and points for consideration when executing tasks required for vehicle maintenance and inspection. This analysis designates three distinct locations for vehicle maintenance, two distinct working areas on the vehicle itself, and designates tasks to be performed in these areas as high risk or low risk.

- Any fall, even a fall on the same level, can result in disability or death; but for the purpose of this analysis, fall severity shall be generalized, based on potential fall distance.
  - Falls of four feet or less are considered relatively minor.
  - Falls from 4-10 feet are considered moderate falls that likely would result in bone fractures, severe contusions, and a lost-time injury.
  - Falls of greater than 10 feet are considered severe and may result in severe injury including partial disability or death.
  - Falls from any height where there is potential of falling onto hazardous conditions (e.g., machinery) are considered severe.

Mitigation Method. Fall protection and fall protection methods are parts of the overall hazard analysis and prevention plan for a vehicle maintenance and/or inspection evolution. Location of the vehicle or potential fall exposure, nature of the task, work area of the vehicle or working platform, and consideration for other hazard potentials that may be introduced with the use of fall protection methods, shall be considered for each task.

- For example, moving vehicles in order to make space for fall protection equipment, especially in confined and/or areas with high density of other vehicles and equipment, may create additional hazards that outweigh those of the fall hazard.
- Consideration must be given to all hazards involved with the execution of a particular maintenance or inspection task to determine the best and safest course of action to protect personnel as well as equipment.
- Hazard Analysis. Exposures to fall hazards are considered as the product of frequency of exposure, proximity to fall hazard, duration of exposure, security of area, and the environmental impact. These factors provide the overall probability of the fall.
- The hazard assessments provided in this document are displayed using tables which quantify fall hazards based on severity and probability of a hazard becoming a mishap
- Where multiple fall hazards exist during a particular work task, the one with the highest associated risk (worst case) shall be utilized to determine the proper fall hazard controls to use during the entire work task.
- This assessment is not all-inclusive of all hazards or conditions that may arise, but is a guideline for supervisors and workers to execute operational risk management, to protect personnel and materiel.

Fall Hazard Areas:
- Tasks and Risk Management hazard assessments
  - Tanker
    - Fall Hazard: Height of Total Exposure is approximately 10 feet from the top of tanker to the ground. High-Risk Tasks include:
    - Loading fuel through the manhole located at the top of the tank.
    - Removal, installation, and/or maintenance of V14 pilot valve.
    - Removal, installation, and/or maintenance of V19 jet sensor valve.
    - Removal, installation, and/or maintenance of V10 bottom load valve.
    - Removal, installation, and/or maintenance of fuel level sensor.
    - Low-Risk Tasks include:
    - Pre-operation inspections to include daily and turnaround maintenance pre-operation inspections.
    - Removal, installation, and/or maintenance of DLPG/VNPG gas tubing
    - Removal, installation, and/or maintenance of auxiliary pump junction box.
    - Risk: Steps integrated into vehicle provide access

to entry hatches and associated components. Steps may be inadequate to maintain balance and mitigate risk of falling while attempting to perform vehicle maintenance and inspection. The risk of falling increases significantly with less than three points of contact between the vehicle and the person performing vehicle maintenance or inspection.

- ◆ Hazard assessment:

HIGH-RISK TASK RAC MATRIX

Hazard Severity	Probability			
	A	B	C	D
I	1	1	2	3
II	1	2	3	4
III	2	3	4	5
IV	3	4	5	5

LOW-RISK TASK RAC MATRIX

Hazard Severity	Probability			
	A	B	C	D
I	1	1	2	3
II	1	2	3	4
III	2	3	4	5
IV	3	4	5	5

- ◆ Mitigations:
- ◆ Fall protection stand with appropriate PPE.
- ◆ Use of integrated vehicle steps with appropriate PPE.
- ◆ Deliberate Risk Management/Administrative controls. Some examples can include:
- Consider specific part of vehicle used to climb onto vehicle and determine whether or not a safer alternate exists.
- Check weather conditions and postpone vehicle climb if necessary.
- Ensure vehicle surfaces are free of slippery substances.
- Assure that boot soles are in good condition.
- Assure the climbing personnel maintain a minimum of three points of contact at all times.
- Don head protection.
- Tractor
  - ◆ Fall Hazard: Height of Total Fall Exposure is approximately four feet to seven from the ground. This height is dependent upon the location, tire diameter and the suspension height.
  - ◆ High-Risk Tasks to be performed:
  - ◆ Machine gun ring removal/installation.
  - ◆ Tire davit winch and cable removal/repair/installation.
  - ◆ Low-Risk Tasks include:
  - ◆ Pre-operation inspections, to include daily and turnaround maintenance pre-operation inspections.
  - ◆ Inspection of fifth wheel mounting for bent, worn, or broken parts.
  - ◆ Inspection and maintenance on tractor engine.

- ◆ Risk: Steps integrated into vehicles provide access to the engine area, transmission, tanker, and associated components. Steps and working surfaces may be inadequate to maintain balance and mitigate risk of falling while attempting to perform vehicle maintenance and inspection in these areas. The risk of falling increases significantly with less than three points of contact between the vehicle and the person performing vehicle maintenance or inspection.
- ◆ Hazard assessment:

HIGH-RISK TASK RAC MATRIX

Hazard Severity	Probability			
	A	B	C	D
I	1	1	2	3
II	1	2	3	4
III	2	3	4	5
IV	3	4	5	5

LOW-RISK TASK RAC MATRIX

Hazard Severity	Probability			
	A	B	C	D
I	1	1	2	3
II	1	2	3	4
III	2	3	4	5
IV	3	4	5	5

- ◆ Mitigations:
- ◆ Fall protection stand with appropriate PPE.
- ◆ Use of integrated vehicle steps with appropriate PPE.
- ◆ Deliberate RM/Administrative controls. Some examples can include:
- Consider specific part of vehicle used to climb onto vehicle and determine whether or not a safer alternate exists.
- Check weather conditions and during inclement weather postpone vehicle climb if necessary.
- Ensure vehicle surfaces are free of slippery substances.
- Assure boot soles are in good condition.
- Assure that climbing personnel maintain minimum three points of contact at all times.
- Don head protection.
- Additional Hazard analysis for consideration:
- Vehicle movement
- Hazard: Vehicle Movement – The M978 tanker has a gross combined weight rating of 100,000 pounds. Vehicle moves can be highly dynamic evolutions that put personnel at risk of being fatally crushed or injured.
  - ◆ Where vehicles are parked in close quarters, the hazards associated are increased. Close quarters also make the vehicle movement much more technically demanding for personnel operating the vehicle being moved.
  - ◆ Ground guide personnel may be unable to remain safely out of the path of the vehicle.

- Hazard assessment:

Hazard Severity	Probability			
	A	B	C	D
I	1	1	2	3
II	1	2	3	4
III	2	3	4	5
IV	3	4	5	5

- Mitigations:
  - Use of appropriate, properly trained personnel.
  - Use of deliberate risk management prior to moving vehicle.
- Movement of fall protection equipment.
  - Hazard: M978 vehicles transport flammable liquids. Combustible vapors could potentially be produced and released from the vehicle. Movement of fall protection stands could result in vehicle damage.

◆ Hazard assessment:

Hazard Severity	Probability			
	A	B	C	D
I	1	1	2	3
II	1	2	3	4
III	2	3	4	5
IV	3	4	5	5

- Mitigations:
  - Use of appropriate, properly trained personnel to include spotters to mitigate contact between fall protection stands and fuel tanker.
  - Use of deliberate RM prior to moving equipment.

The elements contained in this Fall Hazard Survey Report have been reviewed and approved.

_____________(sign)_____________ _______________

Fall Protection Program Manager   Date

_____________(sign) _____________ _______________

Competent Person for Fall Protection   Date

Appendix N
GROUND MAINTENANCE FALL PROTECTION AND PREVENTION PLAN

- Description of fall hazards that may be encountered at the workplace by end-users during performance of their work. The M978 has two primary work areas that may expose personnel to fall hazards, as the working surfaces are greater than four feet above ground level. The working area behind the cab shall be generalized and referred to as the engine area. This area potentially exposes personnel to a fall hazard of four to seven feet. The trailer area shall be generalized as the tanker area and potentially exposes personnel to a ten-foot fall hazard.
- Type of fall protection/fall prevention methods or systems used for every phase of work. Military vehicle maintenance and inspection is inherently dynamic and fraught with potential hazards. 29CFR 1910 requires fall protection when working four feet or more above lower levels to be mitigated by one or more of five control measures listed in Chapter 6. Additionally, Federal regulation mandates this requirement, except where employers can demonstrate that such fall protection systems are infeasible or would create a greater hazard.
  - Fall protection control measures are parts of the overall hazard analysis and fall protection and prevention plan for vehicle maintenance and/or inspection. Location of the vehicle or potential fall exposure, nature of the task, work area of the vehicle or working platform, and consideration for other hazard potentials that may be introduced with the use of fall protection control measures, shall be considered for each task.
  - For example, moving a vehicle in order to make space for fall protection equipment to execute a low risk task, especially in confined and/or areas with high density of other vehicles and equipment, adds additional hazards that may outweigh those of the unprotected fall hazard.
  - Consideration must be given to all hazards involved with the execution of a particular maintenance or inspection task to determine the best and safest course of action.
  - Fall Hazards shall be mitigated according to the hierarchy of controls.
  - The overall approach to fall prevention at Unit Name shall be in the form of prevention, engineering controls, and administrative procedures. As there are no anchor

points or sufficient height above ground for proper employment of a fall-arrest system on the M978 vehicle without the use of external equipment or an engineering change to the hangar, the use of fall-arrest system is not practicable at this time for Unit Name.

- Feasibility of a fall-arrest system for applications within the maintenance area shall be explored in the future at Unit Name.
- Unit Name has fall protection stands that provide both fall prevention, on the backside of stands with handrails, and engineering controls, on the vehicle side of the stands, to minimize the potential for falls. The unit also has B-1 and B-4 work stands, phase stands, and ladders that provide some degree of fall prevention and engineering controls.
- There are some locations where the use of stands is not practicable, due to overall hazard analysis, or mission requirements. In these situations or areas, administrative controls shall be utilized. The primary administrative controls utilized at Command Name are the use of qualified climbers, and deliberate and time critical RM. Three primary vehicle locations shall be addressed in this plan: parking area, maintenance bay, and wash rack.
- Field Maintenance: Fall protection equipment and systems generally are not utilized during operations. Use of fall protection equipment in this environment may cause a potential for greater hazards than the fall hazard itself.
- Fall protection equipment normally should not be utilized in a parking area at Unit Name unless a risk management assessment indicates that the presence of the equipment does not create a larger overall hazard to personnel than the risk of a fall.
- Maintenance should be performed in a maintenance area whenever feasible.
- Vehicle inspection and maintenance in parking areas shall require the use of deliberate and time critical risk management and restricting access to qualified climbers in order to minimize the risk of a fall.
- In the consideration of using maintenance stands and/or fall-arrest systems where fall potential might be increased by moving vehicle into a maintenance area, RM alsoshould be applied.
- Consideration should be given to moving vehicle requiring high risk maintenance actions into the maintenance area in order to utilize maintenance stands and/or fall-arrest systems to prevent a fall.

- Wash rack: Maintenance stands with guards or arrest systems shall be used for all vehicle washing procedures where maintenance personnel are on the vehicle greater than four feet above ground level.
- Other maintenance and inspection tasks executed at the wash rack not associated with vehicle washing shall be executed using the same fall protection measures as normally would be used in the maintenance area.
- Maintenance Area:
- Trailer. Due to the nature of the steps and increased potential for falls from greater heights, fall protection maintenance stands (with guardrails at the entire periphery) shall be used to the maximum extent practicable on all maintenance actions top of the tank where the potential for a fall of ten feet exists. B-1, B-4, phase, and/or fall protection stands shall be configured and positioned properly to mitigate the fall hazard.
- Should the use of a stand be deemed impracticable; a Competent Person for Fall Protection, and Maintenance Control shall approve deviations and determine alternative fall protection methods, case-by-case. Nonetheless, railed maintenance stands, arrest systems, or other fall protection measures should be used, except in rare exceptions.
- Ladders alsomay be used for tasks determined low-risk by the fall hazard assessment, as long as they are used and set up properly. Personnel utilizing the ladder shall maintain three points of contact with the ladder at all times, and the ladder itself shall be positioned with four points of contact at all times.
- Tractor engine area. The engine area has a maximum fall exposure of seven feet. Risk assessment for the specific task to be executed can be found in the hazard assessment (see Fall Hazard Survey).
- Fall protection stands shall be used to the maximum extent practicable on all maintenance actions in the hydraulics bay area, where the potential for falling four feet or greater exists.
- Where B-1, B-4, phase, and/or fall protection stands are used, they shall be configured and positioned properly to mitigate the fall hazard, and to prevent the creation of additional fall hazards which may be caused by improperly configured stands.
- Should the use of a stand be deemed impracticable for a high-risk task as defined in the hazard assessment; a Competent Person for Fall

Protection and a Maintenance Control representative shall approve deviations; and they shall determine alternative fall protection methods, case-by-case.

- ◆ Should the use of a stand be deemed infeasible for a low-risk task as defined in the hazard assessment (Fall Hazard Survey); the end-user, under the direction of a Competent Person for Fall Protection, may utilize administrative controls, which shall include deliberate risk management prior to execution of the task.

- The use of administrative controls as the sole fall protection methodology is authorized only during the rare occasions when the movement of, setup, and use of a fall protection stand or arrest system creates a greater hazard than the hazards associated with the low-risk maintenance action to be performed. For example, the hazard of moving maintenance or fall-arrest system stands into a parking area may outweigh a low-risk fall hazard.

- Normally, engineering controls (maintenance stands) or other controls to prevent a fall from occurring (fall-arrest system) always should be used.

- Training requirements for personnel exposed to fall hazards. Unit Name civilians and military personnel who use fall protection equipment (stands); and other Army personnel involved in the fall protection program; shall be trained in accordance with OSHA, in order to recognize, evaluate, and control fall hazards.

  - Initial Training. Command personnel exposed to fall hazards shall complete the following training:
    - ◆ Fall hazard awareness training: general fall hazard awareness training shall be provided by either a Competent Person for Fall Protection or by completion of an approved general fall hazard awareness training.
    - ◆ Personnel required to climb vehicles shall be qualified climbers.
    - ◆ Qualified climbers shall be trained and knowledgeable regarding what surfaces are or are not able to support a climber and shall observe vehicle restrictions. Training shall include:
  - Requirement for head protection
  - Identification of areas on vehicle authorized for use as a step
  - Checking to ensure that steps are cleaned of slippery substances
  - Methods to maintain three points of contact on vehicle at all times

- ◆ Qualified climbers shall rely on experience, proper training, and the use of time-critical risk management to minimize their risks of a fall.
- ◆ End-users of Fall-arrest Systems also shall undergo "Hands-On" training on the specific fall-arrest system equipment used at the command. This training must be performed by a Competent Person for Fall Protection. End-user training shall include training on:
  - Safe use of equipment
  - Proper application of equipment
  - Equipment limits
  - Estimation of fall distances and clearance
  - Proper anchoring and tie off techniques
  - Methods of inspection, storage, care, and maintenance
  - Applicable regulations
  - Recognition of fall-hazard deficiencies
  - Site-specific procedures
  - Retraining. Retraining in relevant topics shall be provided to the end-user and/or qualified climbers where:
    - ◆ The end-user or qualified climber has been observed using fall protection equipment in an unsafe manner.
    - ◆ The end-user or qualified climber has been involved in a mishap or a near-miss incident.
    - ◆ The end-user or qualified climber has received an evaluation that reveals that he or she is not using the fall protection equipment properly.
    - ◆ The end-user or qualified climber is assigned a different type of fall protection equipment.
    - ◆ A condition in the workplace changes in a manner that could affect the safe use of the fall protection equipment that the end-user is to utilize.
  - Refresher training. Personnel exposed to fall hazards shall receive refresher training on the safe use of fall protection equipment at a minimum of every two years.

- Type of fall protection equipment and systems provided to the Soldiers/employees that might be exposed to fall hazards.
  - Fall Protection Stands:
    - ◆ Vehicle Maintenance Platform Type B-1
    - ◆ Vehicle Maintenance Platform Type B-4
    - ◆ West Coast Weld Tech H-60 Helicopter Maintenance Platform
    - ◆ Metallic Ladder Mobile Work Platform SCS 8460 and SCS 9636-44

- Ladders: Ladders are not technically fall protection equipment though they may be used to ascend working or inspection areas at heights when the use of fall protection equipment is not feasible or not available.
- Fall-arrest Systems
  - When used as part of the fall protection program, fall-arrest systems (harnesses, lanyards, and associated hardware) shall be inspected annually or following the manufacturer's instructions whichever is more stringent.
  - Fall-arrest systems shall be inspected annually by a Competent Person for Fall Protection and inspection documentation shall be maintained by the FPPM for a period of at least five years.
- Current Competent Persons
  - Competent Person for Fall Protection: As designated by the commander and trained in accordance OSHA standards.
  - Competent Person for Equipment Inspection: As designated by the commander.
- Fall protection equipment and instructions for assembly, disassembly, storage, maintenance, and care. B-1 and B-4 maintenance railed platforms are considered support equipment and shall be assembled, disassembled, and maintained by the support equipment division.
  - Pre-operational inspections shall be conducted prior to each use by the end-user in accordance with manufacturer's instructions.

- Phase and other maintenance stands not under the purview of the support equipment division shall be assembled, disassembled, and maintained per manufacturer instructions at a maximum one-year interval by a Competent Person for Fall Protection or a Competent Person for Equipment Inspection.
- Maintenance stands shall be inspected by the end-user using the Unit Name (local) pre-operational checklist. Fall protection maintenance stands shall be stored in accordance with manufacturers' instructions.
- Ladders: Ladders shall be inspected by the end-user prior to each use, and inspected at maximum one-year intervals by a Competent Person for Fall Protection or a Competent Person for Equipment Inspection.

• Rescue plan and procedures. Unit Name does not currently employ the use of fall-arrest systems or equipment. Should this type of equipment be utilized in the future by Unit Name personnel, the FPPM shall coordinate with the Competent Person for Fall Protection support to create a rescue plan which meets the requirements in 29CFR1910. The rescue plan shall then be incorporated into the command Fall Protection Program.

• The elements contained in this Fall Prevention and Protection Plan have been reviewed and approved.

______________________________   ___________
Unit Fall Protection Program Manager   Date

______________________________   ___________
Competent Person for Fall Protection   Date

Appendix O
GLOSSARY

Activation Distance: The distance traveled by fall arrestor or the amount of line played out by self-retracting lanyard from the onset of a fall to the point where the system locksoff.

Active Fall Protection System: A fall protection system that requires end-users to wear or use fall protection equipment.

Anchorage: A secure point of attachment for lifelines, lanyards, or deceleration devices.

Anchorage Connector: A component or subsystem by which fall protection or rescue equipment is secured to the anchorage. This can include a steel cable sling, tie-off adopter (anchor strap), load-rated eyebolt, tripod, davit arm, or any other device designed to suspend human loads and capable of withstanding forces generated by a fall.

Anchorage System: A combination of anchorage and anchorage connector.

Arresting Distance: The total vertical distance required to arrest a fall. Includes activation and deceleration distance.

Arresting distance does not include free-fall distance.
Arresting Force: Force exerted on a worker or test weight, when a fall protection system stops the fall. The amount usually expresses the peak force experienced during a fall.

Assigned Safety Person (Spotter): A Soldier/employee assigned to periodically check (at least every 5 minutes) visually or verbally to assure that an end-user has not fallen and is suspended in his/her harness. This assigned safety person shall have the ability to make quick contact with the jurisdictional public/Government-emergency response agency. This is also known as the "Buddy System."

Assisted Rescue: A planned means of rescue, requiring the assistance of others.

Authorized Person: See the definition of end-user.

Authorized Rescuer: A person who is trained on rescue procedures and assigned by the command to rescue an end-user who may require rescue.

Automatic Controlled Descent Device: A personal lowering device or mechanism that once engaged will automatically control pay-out speed of line or descent speed under load, self-adjusting for a person's weight and operating by gravity. Some automatic controlled descent devices have self-retracting lanyard capability.

Available Clearance: The distance from the walking-working surface or platform to the nearest obstruction that the end-user might contact during a fall.

Back-Strap: A strap located on the back of a full-body harness that connects between the straps below the dorsal location and above the waist, which is intended to keep the body from exiting the rear of the harness.

Body Belt: A strap with means both for securing it about the waist and for attaching it to a lanyard, lifeline, or deceleration device.

Body Harness: Means of configuration of connected straps secured about the Soldier/employee in a manner that will distribute the arresting forces over at least the upper thighs, waist, shoulders, chest and pelvis, with means for attaching a lanyard to other components of the personnel fall arrest system. A full-body harness is the only body support device allowed by OSHA or ANSI when a free-fall distance exceeds two feet.

Boatswain (Bos'n) Chair: A single-point adjustable suspension scaffold consisting of a seat or strap designed to support one Soldier/employee in a sitting position. The seat is made of a plywood or strap independently suspended from an anchorage and the Soldier/employee using full-body harness is attached to a separate lanyard or lifeline attached to an independent anchorage may sit to help alleviate the pooling of blood in the legs.

Brake Bar Rack: A series of smooth bars connected together in parallel in which a synthetic rope is intertwined so the friction of the rope against the bars controls the descent of a lowering device (often used in a rope rescue system).

Buckle: A connector used for attaching the strap or webbing segments or to themselves.

Capacity: The maximum weight that a component, system, or subsystem is designed to hold. This includes the combined weight of the user, clothing, tools and other objects carried by the end-user.

Cable Grab: See fall arrestor

Carabiner: A connector component generally consisting of an oval or trapezoidal shaped body with a closed gate or similar arrangement. Only self-locking carabiners are acceptable for use.

Certified Anchorage: A fall protection or rescue anchorage that a qualified person certifies to be capable of supporting the potential forces that could be encountered in the process of arresting a fall.

Clearance: The distance from a specified reference point, such as the working platform or anchorage of a fall-arrest system, to the lower level that a worker might encounter during a fall.

Clearance Requirement: The distance below the end-users that must be clear of obstructions in order to ensure that the end-user does not encounter any object or obstruction during a fall.

Cleat: A ladder crosspiece of rectangular cross section placed on edge upon which a person may step while ascending or descending a ladder.

Climbing Ladder Fall Arrest System: See "Ladder Climbing (Safety) Device."

Competent Person (CP) for Fall Protection: A person designated by the command to be responsible for the immediate supervision, implementation and monitoring of the fall protection program, who through training knowledge and expertise is capable of identifying, evaluating and addressing existing and potential fall hazards and in the application and use of personal fall arrest and rescue system or any component thereof, and who has the authority to take prompt corrective measures to eliminate or control the hazards of falling.

Competent Rescuer: An individual designated by the employer who by training, knowledge and experience is capable of the implementation, supervision and monitoring of the employer's fall protection rescue program.

Connector: A device that is used to couple (connect) parts of a personal fall arrest system or positioning device system together.

Connecting Means: The method to connect body support to an anchorage, such as a lanyard, snap-hook or a carabiner for the purpose of providing protected mobility for an elevated work task.

Connecting Subsystem: An assembly, including the necessary connectors, comprised of components, subsystems or both, between the anchorage connector and the D-ring of the body harness.

Continuous Fall Protection: One or more fall protection systems that provide fall protection without interruption.

Controlled Access Zone (CAZ): A zone to restrict access to unprotected edge work. The CAZ is bound by a control line and should run the full length of the unprotected edge and connect on each side to a guardrail or wall. The control line can be made of rope, wire, tape, or equivalent material and shall be supported by stanchions and marked with a highly visible material. Controlled Access Zones are not allowed on USACE projects

Deceleration Device: Any mechanism, such as a rope, grab, ripstitch lanyard, specially woven lanyard, tearing lanyard, deforming lanyard, or automatic self-retracting lifeline/lanyard, which serves to dissipate a substantial amount of energy during a fall arrest, or otherwise limits the energy imposed on a Soldier/employee during fall arrest.

Deceleration Distance: The vertical distance measured between the location of the user's fall arrest attachment point (dorsal D-ring) at the onset of fall arrest forces during a fall, and after the fall arrest attachment point comes to a complete stop. Is the additional vertical distance a falling Soldier/employee travels, excluding lifeline elongation and free-fall distance, before stopping, from the point at which the deceleration device begins to operate?

Descent Controller: A device designed to be used by one worker for personal descent to lower another person from an elevation. Descent control may be used for egress, positioning or both.

Designated Area Method: A distinct portion of a walking-working surface delineated by a perimeter warning line in which temporary work may be performed without additional fall protection. The designated area method is only used for general industry work.

D-ring: A connector used integrally in a harness as an attachment element for fall arrest. It is alsoused in lanyards, energy absorbers, lifelines, and anchorage connectors as an integral connector.

Dorsal: A location on a full-body harness that falls approximately between the user's shoulder blades.

Double-Cleat Ladder: A ladder with a center rail to allow simultaneous two-way traffic for Soldiers/employees ascending or descending.

End-User of Fall Protection (Authorized Person): A person who has been trained in the use of assigned fall protection equipment, including handsoon training and practical demonstrations in a typical fall hazard situation, and uses personal fall arrest or restraint/positioning equipment while performing work assignments at heights.

Energy (Shock) Absorber: A component whose primary function is to dissipate energy and limit deceleration forces that the system imposes on the body and the anchorage system during fall arrest.

Engineered Anchor: An anchorage designed and approved by a qualified person.

Energy Absorber, Horizontal Lifeline: An energy absorber that is attached to one of the end anchorages or anchorage connectors of a horizontal lifeline subsystem.

Energy Absorber, Personal: An energy absorber that is attached to a harness.

Energy Absorber, Single Anchor Vertical Lifeline: An energy absorber that is attached to the top anchorage or anchorage connector of a single anchor vertical lifeline subsystem.

Evacuation Harness: A component for rescue purposes consisting of elements designed and constructed so that the rescue subject is securely held during the rescue process. Evacuation harness is a special harness.

Failure: Load refusal, breakage, or separation of component parts. Load refusal is the point where the ultimate strength is exceeded.

Fall-Arrest System: A combination of equipment and components such as full-body harnesses, lanyards, deceleration devices, anchorages, horizontal or vertical lifelines connected, designed to stop a person from striking a lower level or an obstruction during a fall.

Fall Arrestor: A fall arrest device that locks by either a cam lock (locking arm) or inertia when a free fall is sensed.

It is attached to a worker directly or by a lanyard that slides up or down a fixed or vertical cable or rope lifeline.

Fall Hazard: Any location where a person is exposed to a potential free fall.

Fall Prevention: The elimination and minimization of potential fall hazards, lessening the chance of Soldier/ employee exposure to falls. Any same-level means used to reasonably prevent exposure to a fall hazard; examples of fall prevention are guardrails, walls, floors, and area isolation. Also called a passive fall protection system.

Fall Protection: Action and procedures to effectively protect a worker from fall hazards. Any equipment, device or system that prevents an accidental fall from elevation or that mitigates the effect of such a fall.

Fall Protection Program Manager: A person assigned to be responsible for developing and managing the fall protection program.

Fixed Ladder: A ladder that cannot be readily moved or carried because it is an integral part of a building or structure.

Force Factor: The rati- of peak arresting force using rigid weight compared to a human body having the same weight, both falling under identical conditions.

Forced Rollout: An action by which the gate of a locking snap-hook or carabiner is loaded beyond its design strength forcing it to fail and disengage from the component it was attached to.

Free Fall: The act of falling before a personal-fall-arrest system begins to apply force to arrest a fall.

Free-Fall Distance: The vertical distance from the onset of a fall to a point where a fall-arrest system is activated or engaged. (This is the vertical distance measured from the fall arrest attachment point on the Soldier's/ employee's body harness at the onset of the fall to the point just before the system begins to apply force to arrest the fall. This distance excludes deceleration distance, and lifeline/lanyard elongation which are exerting deceleration forces, but includes any deceleration device slide distance or self-retracting lifeline/lanyard extension before they operate and fall arrest forces occur.)

Frontal d-Ring Attachment: An attachment element affixed to the full-body harness within the vertical sevenoinch sternum (breastbone) area that is designed to withstand dynamic fall arrest, restraint, and postofall suspension forces.

Full-body Harness: See the definition of body harness.

Full body Harness Stretch: The difference between the lowest point on the torso postofall and the lowest point on the torso pre-fall in relation to the attachment element. This accounts for a component of the system stretch out and total fall distance.

Guardrail System: A passive fall protection system of horizontal rails and vertical posts that prevent a person from reaching a fall edge. Guardrail systems typically have a top rail, a mid-rail and posts and toe-board.

Handrail: A rail used to provide Soldiers/ employees with a handhold for support.

Hole: A void or gap 2 inches (5.1 centimeters) or more in the least dimension in a floor, roof, or other walking/working surface.

Horizontal Lifeline (HLL): A fall arrest system that uses a flexible line made from rope, wire rope or synthetic cable that spans horizontally between two end anchorages. The assembly includes the necessary connectors, turnbuckles, inoline energy absorbers, shackles, etc. and may include intermediate anchorages. The system includes fall protection equipment that enables a trained worker to move and safely traverse/work in the horizontal plane.

Horizontal Lifeline Subsystem: An assembly, including the necessary connectors, comprised of a horizontal lifeline component and, optionally, of: a) An energy absorber component or, b) a lifeline tensioner component, or both. This subsystem is normally attached at each end to an anchorage or anchorage connector and may contain one or more intermediate anchorages.

The end anchorages have the same elevation.

Horizontal Track System: A form of rigid rail system that typically encloses a trolley inside a formed channel or track.

Hybrid Component: An integral assembly of elements or components, or both, intended to perform more than one function in the system.

Ladder Climbing (Safety) Device: A device or climbing sleeve connected to the front D-ring on the climber's full-body harness that slides up or down a rigid rail or cable. Should a fall occur, the device is designed to lock by inertia or cam action to arrest the fall.

Lanyard: A flexible line of rope, wire rope, or strap that generally has a connector at each end for connecting the body belt or body harness to a deceleration device, lifeline, or anchorage.

Leading Edge: The edge of a floor, roof, or formwork for a floor or other walking/working surface which changes location as additional floor, roof, decking, or formwork sections are placed, formed, or constructed.

Lifeline: A component consisting of a flexible line for connection to an anchorage at one end to hang vertically (vertical lifeline), or for connection to anchorages at both ends to stretch horizontally (horizontal lifeline), that serves as a means for connecting other components of a personal fall arrest system to an anchorage.

Lifeline Tensioner: A device, such as a turnbuckle, to tauten a horizontal lifeline or a weight to tension a vertical lifeline.

Load-bearing Straps: Straps as part of the full-body harness through which load is transmitted during a fall or under normal use.

Load Refusal: The point where the structural members lose their ability to carry the load.

Low Slope Roof: A roof having a slope less than or equal to 4 in 12 (vertical to horizontal).

Man Overboard Plan: A man overboard plan is an emergency plan for rescuing personnel if they accidentally fall in the water.

Manual Descent Controlled Device: A load lowering device or mechanism that once engaged requires manual attention to control payout speed of line or descent speed under load.

Manual Fall Arrester (Manual Rope Grab): A fall arrester that will remain locked where it has been positioned on a VLL until deliberately repositioned by a worker.

Marking: Any sign, label, stencil, plate or the like containing information or guidance.

Maximum Arresting Force (MAF): The peak force exerted on the body or test weight when a fall protection system arrests or stops a fall.

Maximum Arrest Load (MAL): The peak force applied to an anchorage by an active fall protection system when arresting a fall.

Military Unique: Operations conducted under the guidance of Army Regulations due to the unique nature of the mission. Not all aspects of OSHA will apply, however they should be complied with where feasible.

Non-Certified Fall Protection Anchorages: An unquestionably strong anchorage that a Competent Person judges to be capable of supporting the predetermined anchorage strength as prescribed by OSHA Standards and ANSI/ASSE Fall Protection Code. Non-certified anchorages are used either for fall arrest, work positioning, travel restraint or rescue.

Opening: A gap or void 30 inches (76 centimeters) or more high and 18 inches (46 centimeters) or more wide, in a wall or partition through which Soldiers/employees can fall to a lower level.

Orthostatic Intolerance (Suspension Trauma): The development of symptoms as a result of suspension in a full-body harness, such as light-headedness, palpitations, tremulousness, poor concentration, fatigue, nausea, dizziness, headache, sweating, weakness, and occasionally fainting and unconsciousness.

Passive Fall Protection System: A system that does not require a worker to use or wear personal fall arrest equipment. Examples include safety nets, guardrails, parapet walls, etc.

Personal Fall Arrest System: A system including but not limited to an anchorage, connectors, and a body harness used to arrest a Soldier/employee in a fall from a working level.

Point of Access: All areas used by Soldiers/employees for work-related passage from one area or level to another.

Portable Ladder: A ladder that can be readily moved or carried.

Positioning Device System: A body belt or body harness system rigged to allow a Soldier/employee to be supported on an elevated vertical surface, such as a wall, and work with both hands free while leaning backwards.

Positioning Lanyard: A lanyard used to transfer forces from a body support to an anchorage or anchorage connector in a positioning system.

Pre-Incident Plan: A formal written plan, prepared jointly by the host unit and the fire emergency responders, containing factors that need to be evaluated when assessing the potential situations that could affect a facility during emergency conditions.

Primary System: In fall protection terminology, the main mechanism that allows a worker to maintain his or her desired position.

Qualified Person (QP) for Fall Protection: A person with a recognized engineering degree or professional certificate and with extensive knowledge training and experience in fall protection and rescue filed, who is capable of performing design, analysis, evaluation, and specifications of fall protection and rescue systems and equipment.

Rescue: Process of evacuating a person or persons to a safe location where they also may receive medical attention.

Rescue Cradle: A cradle made of synthetic material with polyester webbing with integrated steel rings attachable to flexible line of rope or strap via a carabiner, used to lower injured personnel from heights.

Rescue Ladder: A flexible ladder with rigid rungs and either synthetic webbing or wire rope side rails which can be temporally hung next to the end-user working at heights, or can be lowered down to an end-user suspended in a harness to allow him to climb back up to the working surface (or at least stand on the ladder while waiting rescue allowing the necessary circulation of the blood to the entire body while an assisted rescue is being commenced).

Rescue Lanyard: A component consisting of flexible line of rope or strap, which generally has a connector at each end for connecting the body support to components of a rescue system. A rescue lanyard is a special lanyard.

Rescue Plan (Fall Arrest): A written plan that describes the rescue method and procedures to be used to rescue an end-user of fall protection who may have fallen from a height and be suspended in a full-body harness. The suspended worker may have been injured or incapacitated prior to, or as the result of the fall.

Restraint System: A combination of devices designed to restrain an end-user from reaching an exposed fall hazard. The system consists of a full-body harness that can be secured around a worker and attached to a load-bearing anchorage in order to restrict travel and limit fall hazards. The strap can be single or multiple.

Rigid Anchorage Subsystem: An anchorage system, such as a rigid rail or trolley system or a single point of attachment that does not appreciably deflect, deform, or stretch when a fall-arrest impact occurs.

Rigid Rail System: A fall protection system that uses one or more trolleys on a horizontal track (often an I-beam or slotted tube).

Rigging: The process of building a system to move or stabilize a load or the system itself.

Riser Height: The vertical distance from the top of a tread or platform/landing to the top of the next higher tread or platform/landing.

Rollout: An action by which a snap-hook or carabiner unintentionally disengages from another connector or object to which it is attached.

Rope Access: A system consisting of two lifelines independently anchored at the top to protect the authorized person from falling. The ropes directly suspend the person. The technique is used on buildings, bridges, and other structures for conducting inspection, cleaning, and painting.

Rope (or Strap) Adjuster: A mechanical means of readily moving a vertical line attachment or changing the position of an intermediate anchorage device between an anchorage (connector) and a body support while loaded with the authorized person's weight or partial weight while leaning.

Rope Grab: A deceleration device that travels on a lifeline and automatically, by friction, engages the lifeline and locks to arrest a fall. See Fall Arrester.

Rope, Synthetic: A construction of bundled manmade yarns, fibers, or filaments forming a strong flexible line.

Rope, Wire: A plurality of drawn wires forming strands laid helically over an axis or core.

Runway:
1. A passageway for person, elevated above the surrounding floor or ground level, such as a foot walk along shafting or a walkway between buildings.
2. Elevated crane rails upon which an overhead electric crane travels.

Safety Margin: A clearance factor of safety defined as the distance between the lowest extremity of the worker's body at fall arrest and the highest obstruction the worker might otherwise make contact with during a fall.

Safety Monitoring System: A safety system in which a Competent Person is responsible for recognizing and warning Soldiers/employees of fall hazards.

Safety Strap/Relief Step Strap: A coiled strap in an attached pouch to the lanyard which is manually deployed after a fall, and allows the end-user to insert one foot (or two feet depending on the style) into the loop step and stand allowing the necessary circulation of blood to the entire body while an assisted rescue is being commenced.

Safety Net System: A horizontal or semi horizontal cantilever-style barrier that uses netting system to stop falling workers before they make contact with a lower level or obstruction.

Sag: The distance the wire rope or synthetic cable of a horizontal lifeline deviates from the horizontal plane established by the end anchorage. This is defined by the line between two anchorages and measuring downward at the mid-point of the wire rope or cable.

Secondary Fall Protection System: One or more means of fall protection, as defined by these standards, configured as a supplement or as backup to protect a worker from a potential fall if the primary system fails.

Self-Retracting Device (SRD): A device that contains a drum wound line that automatically locks at the onset of a fall to arrest the user, but that pays out from and automatically retracts onto the drum during normal movement of the person to whom the line is attached. After onset of a fall, the device automatically locks the drum and arrests the fall. Self-retracting devices include self-retracting lanyards (SRL's), self-retracting lanyards with integral rescue capability (SRL-R's), and self-retracting lanyards with leading edge capability (SRL-LE's) and, hybrid combinations of these.

Self-Retracting Lanyard (SRL): A deceleration device containing a drum-wound line which can be slowly extracted from, or retracted onto, the drum under slight tension during normal Soldier/employee movement, and which, after onset of a fall, automatically locks the drum and arrests the fall.

Self-Retracting Lanyard with Integral Rescue Capability: A device meeting ANSI/ASSE Z359 Fall Protection Code/Standards definition for self-retracting lanyard and including integral means for assisted-rescue via raising or lowering the rescue subject.

Self-Retracting Lanyard with Leading Edge Capability (SRL-LE): A self-retracting device used for horizontal applications which is mounted or anchored at "foot" level and where there is the possibility of free fall. The device includes integral means to withstand impact loading of the line constituent with a sharp or abrasive edge during fall arrest and for controlling fall arrest forces on the user. The device can also be used for vertical applications where it is mounted overhead.

Seat Sling: A seat sling designed for attachment to a full-body harness that is designed so that a worker may sit for a short period of time without pooling of blood in the legs.

Self/Manual Deploying Rescue Ladder: A coiled webbing rescue ladder in a pouch connected to the lanyard or anchorage which either self-deploys during a fall or is manually released by the end-user after a fall, and is left dangling next to the suspended end-user which allows the end-user to climb back up to the anchorage (or at least simply stand in the ladder allowing the necessary circulation of blood to the entire body while an assisted rescue is being commenced).

Self-retracting lifeline/lanyard: A deceleration device containing a drum-wound line which can be slowly extracted from, or retracted onto, the drum under minimal tension during normal Soldier/employee movement and which, after onset of a fall, automatically locks the drum and arrests the fall.

Shock Absorber: See Energy Absorber

Side-step Fixed Ladder: A fixed ladder that requires a person to get off at the top to step to the side of the ladder side rails to reach the landing.

Single Anchor Vertical Lifeline (VLL): A vertically suspended flexible line connected at the upper end for fastening to an overhead anchorage and along which a fall-arrester travels

Single-Cleat Ladder: A ladder consisting of a pair of side rails connected together by cleats, rungs, or steps.

Snap-hook: A connector comprised of a hook-shaped body with a normally closed gate or similar arrangement, which may be opened to permit the hook to receive an object and when it is released, automatically closes to retain the object. Only self-locking (single or double locking) snap-hooks are acceptable for use.

Soft Loop Attachment Element: A non-metallic attachment element of a FBH constructed of synthetic fiber webbing.

Stair Rail System: A vertical barrier erected along the unprotected sides and edges of a stairway to prevent Soldiers/employees from falling to lower levels.

Steep Roof: A roof having a slope greater than 4 in 12 (vertical to horizontal).

Strap: A length of webbing that may be incorporated in a harness, lanyard or other component or subsystem.

Sternal: A location on a full-body harness that falls approximately between the user's chest area.

Suspension: The act of supporting 100% of a user's body weight, including equipment, for the purpose of accessing a work location with one or two points of contact.

Strap, Chest: A harness strap passing generally horizontally across the chest or around the body at chest level with adjustable means for fastening.

Strap, Shoulder: A harness strap that passes from the waist, up the chest, over the shoulder and down the back to the waist. It is connected to the waist strap or thigh straps or sub-pelvic strap or combinations thereof.

Strap, Sub-Pelvic: A full-body harness strap, which passes under the buttocks without passing through the crotch and is designed to transmit, to the sub-pelvic part of the body, forces applied during fall arrest and postofall suspension.

Suspension Trauma (Harness Induced Pathology): Where the body is at rest in a vertical state with the lower body motionless, and as such, blood begins to pool in the lower extremities because the muscles in the legs are not contracting on the veins and helping the blood back to the heart (against gravity). Blood is not properly circulated, the individual's blood pressure drops, the brain does not receive adequate blood flow and unconsciousness follows.

Suspension Trauma Safety Steps/Relief Step Strap: A coiled strap (in a pouch) attached to the harness which is manually deployed after a fall to help prevent the effects of suspension trauma by allowing the end-user to insert one foot (or two feet, depending on the style) into the loop step and stand up allowing the necessary circulation of blood to the entire body, while an assisted rescue is being commenced.

Suspension Work Seat: A seat board with integral body belt, suspension D-rings, and adjustable leg and shoulder straps designed so that a worker may sit for long periods of time without pooling of blood in the legs.

Swing Fall: A pendulum-like motion that can result from moving horizontally away from, or toward, a fixed anchorage and falling. Swing falls generate the same amount of force when falling the same distance vertically. Swing fall has the hazards in both the horizontal direction (swinging into obstruction) and vertical direction (falling onto obstructions or ground).

Synthetic Rope Tackle Block: A load lifting and/or lowering device that does not include a winding or traction drum but use pulleys to achieve a mechanical lifting advantage (often used in a rope rescue system).

Temporary Service Stairway: A stairway where permanent treads and/or landings are to be filled in at a later date.

Testing, Qualification: The controlled application of test conditions to a product specimen randomly selected from the initial production lot, and the recording of observed effects, for the purpose of determining the product's compliance with the requirements of these standards. When the terms "testing" or "tests" are used in the Z359 Standards, those terms shall denote qualification testing or qualification test(s), not developmental or verification testing or test(s) unless otherwise specified.

Testing, Verification: The controlled application of test conditions to a product specimen sampled from ongoing production lots (after qualification testing), and the recording of observed effects, for the purpose of confirming the product's continuing compliance

with the requirements of these standards. Proof load testing is a type of verification testing.

Through Fixed Ladder: A fixed ladder that requires a person getting off at the top to step between the side rails of the ladder to reach the landing.

Toeboard: A low protective barrier that prevents material and equipment from falling to lower levels and which protects personnel from falling.

Total Fall Distance: The maximum distance fallen by the worker using a fall-arrest system between the onset of a fall and the instant when the worker first achieves zer- vertical velocity. Or is the vertical distance fallen by an end-user connected by a fall arrest system to an anchorage measured from the walking/working surface and extending downward to a position after the fall is arrested. The total fall distance includes the sum of the free fall, elongation and deceleration distances of the system.

Travel Restraint: See restraint system

Travel Restraint Lanyard: A lanyard used to transfer forces from a body support to an anchorage or anchorage connector in a travel restraint system.

Tread Depth: The horizontal distance from front to back of a tread, excluding nosing, if any.

Trolley: A mobile anchorage device that travels along a track (horizontal track system), structural beam (rigid rail system), or cable (HLL system).

Unprotected Sides and Edges: Any side or edge (except at entrances to points of access) of a walking/working surface (e.g., floor, roof, ramp, or runway) where there is no wall or guardrail system at least 39 inches (1 meter) high.

Vertical Lifeline (VLL): A vertically suspended flexible line connected at the upper end for fastening to an overhead anchorage and along which a fall arrester travels.

Waist: A location on a FBH corresponding to the area on the body falls typically between the thorax and hips.

Walking/Working Surface: Any surface, whether horizontal or vertical, on which a Soldier/employee walks or works, including but not limited to floors, roofs, ramps, bridges, runways, formwork, and concrete reinforcing steel. Does not include ladders, vehicles, or trailers on which Soldiers/employees must be located to perform their work duties.

Warning Line System: A barrier erected on a roof to warn Soldiers/employees that they are approaching an unprotected roof side or edge and which designates an area in which roofing work may take place without the use of guardrail, body belt, or safety net systems to protect Soldiers/employees in the area.

Webbing: A narrow woven fabric with selvedge edges and continuous filament yarns made from light and heat resistant fibers.

Winch/Hoist: A load lifting and/or lowering device that incorporates a winding drum and means for controlling pay-out and take-up of the line from the drum.

Winch/Hoist Capstan: A load lifting and/or lowering device that incorporates a traction drum and a means for controlling pay-out and take-up of the line from the drum. Device relies on reduction gearing and/or lever principals to achieve a mechanical lifting advantage.

Wire: A single, continuous length of metal with a circular crossosection that is cold-drawn from rod.

Fall Protection in Construction

OSHA 3146-05R 2015

This informational booklet is intended to provide an overview of frequently cited OSHA standards in the construction industry. This publication does not alter or determine compliance responsibilities, which are set forth in OSHA standards and the *Occupational Safety and Health Act*.

Employers and workers in the 27 states and territories that operate their own OSHA-approved workplace safety and health plans should check with their state safety and health agency. Their state may be enforcing standards and other procedures that, while "at least as effective as" Federal OSHA standards, are not always identical to the federal requirements. For more information on states with OSHA-approved state plans, please visit: www.osha.gov/dcsp/osp.

This information will be made available to sensory-impaired individuals upon request. Voice phone: (202) 693-1999; teletypewriter (TTY) number: 1-877-889-5627.

Cover photo: Dona File

Fall Protection in Construction

OSHA 3146-05R 2015

Contents

Why Does OSHA Have a Standard for Fall Protection?

Historically, falls are the leading cause of fatalities in construction, accounting for about one-third of all fatalities in the industry. For example, the Bureau of Labor Statistics reported that there were 291 fatal falls to a lower level in construction in 2013, out of 828 total fatalities.

OSHA recognizes that incidents involving falls are generally complex events, frequently involving a variety of factors. Consequently, the standard for fall protection deals with both the human and equipment-related issues in protecting workers from fall hazards. This publication is intended to help workers and employers better understand the Fall Protection in Construction standard's requirements and the reasons behind them.

What Subpart M – Fall Protection Covers

What is Subpart M?

Subpart M lays out the requirements and criteria for fall protection in construction workplaces. For example, it applies when workers are working at heights of 6 feet or more above a lower level. It also covers protection from falling objects, falls from tripping over or falling through holes, and protection when walking and working around dangerous equipment without regard to height. Subpart M provisions do not apply, however, to workers inspecting, investigating, or assessing workplace conditions prior to the actual start of work or after all construction work has been completed. The provisions of Subpart M can be found in Title 29 Code of Federal Regulations (CFR) Subpart M - Fall Protection, 29 CFR 1926.500, 29 CFR 1926.501, 29 CFR 1926.502, and 29 CFR 1926.503.

What are Employers' Responsibilities to provide Fall Protection?

Initially, employers must assess the workplace to determine if walking or working surfaces have the necessary strength and structural integrity to safely support the workers. Once it is determined that the work surfaces will safely support the work activity, the employer must determine whether fall protection is required (using the requirements set forth in 29 CFR 1926.501) and, if so, select and provide workers with fall protection systems that comply with the criteria found in 29 CFR 1926.502.

When must employers provide Fall Protection? The 6-foot rule.

Subpart M requires the use of fall protection when construction workers are working at heights of 6 feet or greater above a lower level. It applies at heights of less than 6 feet when working near dangerous equipment, for example, working over machinery with open drive belts, pulleys or gears or open vats of degreasing agents or acid.

What construction areas and activities does Subpart M cover?

The standard identifies certain areas and activities where fall protection or falling object protection may be needed. For example, it might require fall protection for a worker who is: on a ramp, runway, or another walkway; at the edge of an excavation; in a hoist area; on a steep roof; on, at, above, or near wall openings; on a walking or working surface with holes (including skylights) or unprotected sides or edges; above dangerous equipment; above a lower level where leading edges are under construction; on the face of formwork and reinforcing steel; or otherwise on a walking or working surface 6 feet or more above a lower level. The standard may also require fall protection where a worker is: constructing a leading edge; performing overhand bricklaying and related work; or engaged in roofing work on low-slope roofs, precast concrete erection, or residential construction. In addition, the standard requires falling object protection when a worker is exposed to falling objects.

What kinds of Fall Protection should employers use?

Generally, fall protection can be provided through the use of guardrail systems, safety net systems, or personal fall arrest systems. OSHA refers to these systems as conventional fall protection. Other systems and methods of fall protection may be used when performing certain activities. For example, when working on formwork, a positioning device system could be used. OSHA encourages employers to select systems that prevent falls of any kind, such as guardrails designed to keep workers from falling over the edge of a building.

Examples of Fall Protection Requirements for Certain Construction Activities

Leading Edges – 29 CFR 1926.501(b)(2)

Each worker constructing a leading edge 6 feet or more above a lower level must be protected by guardrail systems, safety net systems, or personal fall arrest systems. 29 CFR 1926.501(b)(2)(i).

Exception: When the employer can demonstrate that it is infeasible or creates a greater hazard to use these systems, the employer must develop and implement a fall protection plan which meets the requirements of 29 CFR 1926.502(k). See the section below on Fall Protection Plans.

Workers must be protected by guardrail systems, safety net systems, or personal fall arrest systems, *even if they are not engaged in leading edge work*, if they are on a walking or working surface that is 6 feet or more above a level where leading edges are under construction. 29 CFR 1926.501(b)(2)(ii).

Overhand Bricklaying and Related Work – 29 CFR 1926.501(b)(9)

When workers perform overhand bricklaying and related work 6 feet or more above a lower level:

- They must be protected by guardrail systems, safety net systems, or personal fall arrest systems, or
- They must work in a **controlled access zone (CAZ)**.

All workers reaching <u>more</u> than 10 inches below the level of the walking or working surface on which they are working must be protected by a guardrail system, safety net system, or personal fall arrest system.

Roofing Work on Low-Slope Roofs – 29 CFR 1926.501(b)(10)

A low-slope roof has a slope less than or equal to 4 in 12 (vertical to horizontal). When engaged in roofing work on a low-slope roof that has one or more unprotected side or edge 6 feet or more above lower levels, workers must be protected from falling by:

- Guardrail systems,
- Safety net systems,
- Personal fall arrest systems,
- A combination of conventional fall protection systems and warning line systems, or
- A warning line system and a safety monitoring system.

When engaged in roofing work on low-slope roofs 50 feet or less in width, the use of a safety monitoring system without a warning line system is permitted.

Working on Steep Roofs – 29 CFR 1926.501(b)(11)

A steep roof has a slope greater than 4 in 12 (vertical to horizontal). When working on a steep roof that has one or more unprotected side or edge 6 feet or more above lower levels, each worker must be protected by:

- Guardrail systems with toeboards,
- Safety net systems, or
- Personal fall arrest systems.

Residential Construction – 29 CFR 1926.501(b)(13)

Workers engaged in residential construction 6 feet or more above lower levels must be protected by conventional fall protection (i.e., guardrail systems, safety net systems, or personal fall arrest systems) unless another provision in 29 CFR 1926.501(b) provides for an alternative fall protection measure.

Exception: When the employer can demonstrate that it is infeasible or creates a greater hazard to use these systems, the employer must develop and implement a site-specific fall protection plan which meets the requirements of 29 CFR 1926.502(k). See the section on Fall Protection Plans, below.

Note: For purposes of determining the applicability of section 1926.501(b)(13), the term "residential construction" is interpreted as covering construction work that satisfies the following two elements: (1) the end-use of the structure being built must be as a home, i.e., a dwelling; and (2) the structure being built must be constructed using traditional wood frame construction materials and methods. The limited use of structural steel in a predominantly wood-framed home, such as a steel I-beam to help support wood framing, does not disqualify a structure from being considered residential construction. For more information see OSHA's Compliance Guidance for Residential Construction, STD 03-11-002.

Other Walking or Working Surfaces – 29 CFR 1926.501(b)(15)

As a general matter, each worker on a walking or working surface 6 feet or more above a lower level must be protected from falling by a guardrail system, a safety net system, or a personal fall arrest system.

Exceptions: For exceptions to this rule that specify different requirements, see 29 CFR 1926.500(a)(2) and 29 CFR 1926.501(b)(1) through (b)(14).

Guardrail Systems – 29 CFR 1926.502(b)

Guardrail systems are barriers erected to prevent workers from falling to lower levels. If the employer chooses to use guardrail systems to protect workers from falls, the following provisions apply:

- Top rails, or equivalent guardrail system members, must be 42 inches plus or minus 3 inches above the walking or working level. When workers are using stilts, the top edge of the top rail, or equivalent member, must be increased an amount equal to the height of the stilts. 29 CFR 1926.502(b)(1).

- Screens, midrails, mesh, intermediate vertical members, or equivalent intermediate structural members must be installed between the top edge of the guardrail system and the walking or working surface when there are no walls or parapet walls at least 21 inches high. 29 CFR 1926.502(b)(2).

- When midrails are used, they must be installed at a height midway between the top edge of the guardrail system and the walking or working level. 29 CFR 1926.502(b)(2)(i).

Properly installed guardrail system.

- When screens and mesh are used, they must extend from the top rail to the walking or working level and along the entire opening between top rail supports. 29 CFR 1926.502(b)(2)(ii). When necessary, screens and/or mesh must be installed in a manner to prevent a worker from falling underneath.

- When intermediate members (such as balusters) are used between posts, they must not be more than 19 inches apart. 29 CFR 1926.502(b)(2)(iii).

- Other structural members (such as additional midrails and architectural panels) must be installed so that there are no openings in the guardrail system more than 19 inches wide. 29 CFR 1926.502(b)(2)(iv).

- Guardrail systems must be capable of withstanding a force of at least 200 pounds applied within 2 inches of the top edge, in any outward or downward direction, at any point along the top edge. 29 CFR 1926.502(b)(3).

- Midrails, screens, mesh, intermediate vertical members, solid panels, and equivalent structural members must be capable of withstanding a force of at least 150 pounds applied in any downward or outward direction at any point along the midrail or other member. 29 CFR 1926.502(b)(5).

- Guardrail systems must have a surface to protect workers from punctures or lacerations and to prevent clothing from snagging. 29 CFR 1926.502(b)(6).

- The ends of top rails and midrails must not overhang terminal posts, except where an overhang poses no projection hazard. 29 CFR 1926.502(b)(7).

- Steel and plastic banding cannot be used as top rails or midrails. 29 CFR 1926.502(b)(8).

- Top rails and midrails of guardrail systems must have a nominal diameter or thickness of at least 1/4 inch to prevent cuts and lacerations. 29 CFR 1926.502(b)(9).

- If wire rope is used for top rails, it must be flagged at not more than 6-foot intervals with high-visibility material. 29 CFR 1926.502(b)(9).

- When guardrail systems are used at hoisting areas, a chain, gate, or removable guardrail section must be placed across the access opening between guardrail sections during those times when hoisting operations are not taking place. 29 CFR 1926.502(b)(10).

- When guardrail systems are used at holes, they must be set up on all unprotected sides or edges. When a hole is used for the passage of materials, it must not have more than two sides with removable guardrail sections. When the hole is not in use, it must be covered or provided with a guardrail system along all unprotected sides or edges. 29 CFR 1926.502(b)(11) & (12).

- If guardrail systems are used around holes being used as access points (such as ladderways), gates must be used. Alternatively, the point of access must be offset to prevent workers from accidentally walking straight into the hole. 29 CFR 1926.502(b)(13).

- If guardrails are used on ramps and runways, they must be erected on each unprotected side or edge. 29 CFR 1926.502(b)(14).

- Manila, plastic, or synthetic rope used for top rails or midrails must be inspected as frequently as necessary to ensure its strength and stability. 29 CFR 1926.502(b)(15).

Safety Net Systems – 29 CFR 1926.502(c)

When safety nets are used, they must be installed as close as practicable under the walking or working surface on which workers are working and never more than 30 feet below that level. 29 CFR 1926.502(c)(1). When nets are used on bridges, the potential fall area from the walking or working surface to the net must be unobstructed. 29 CFR 1926.502(c)(1). All safety nets must be installed with sufficient clearance underneath to prevent a falling body from hitting the surface or structure below the net. 29 CFR 1926.502(c)(3). If the employer chooses to use nets, the following criteria apply:

Vertical distance from a working level to the horizontal plane of the net	Minimum required horizontal distance from the edge of a working surface to the outer edge of the net
Up to 5 feet	8 feet
More than 5 feet up to 10 feet	10 feet
More than 10 feet	13 feet

Drop-testing is required to ensure that safety nets and safety net installations are working properly. See 29 CFR 1926.502(c)(4)(i) for more details. If an employer can demonstrate that it is unreasonable to perform a drop-test, then the employer or a designated competent person must certify that the net and its installation is in compliance with the standard. See 29 CFR 1926.502(c)(4)(ii) for more details on certification and certification records.

Safety nets used in construction sites.

- Do not use defective nets. Inspect nets at least once a week for wear, damage, or deterioration of components such as net connection points. 29 CFR 1926.502(c)(5).

- Remove materials, tools, and other items as soon as possible from the net and at least before the next work shift. 29 CFR 1926.502(c)(6).

- To work properly, a safety net must have safe openings. Mesh openings must not exceed 36 square inches, and must not be longer than 6 inches on any side. Each opening, measured center-to-center of mesh ropes or webbing, must not exceed 6 inches. 29 CFR 1926.502(c)(7).

- All mesh crossings must be secured to prevent the openings from enlarging. 29 CFR 1926.502(c)(7).

- Use safety net (or section of net) with a border rope possessing a minimum breaking strength of 5,000 pounds. 29 CFR 1926.502(c)(8).

- Do not allow one weak link to compromise a safety net. Use connections between safety net panels that are as strong as integral net components and spaced no more than 6 inches apart. 29 CFR 1926.502(c)(9).

Personal Fall Arrest Systems – 29 CFR 1926.502(d)

A personal fall arrest system is a system used to safely stop (arrest) a worker who is falling from a working level. It consists of an anchorage, connectors, and a body harness. It also may include a lanyard, deceleration device, lifeline, or suitable combinations of these. Under Subpart M, body belts (safety belts) **are prohibited** for use as part of a personal fall arrest system.*

When employers choose to use a personal fall arrest system as a means of worker fall protection they must:

Know the A, B, Cs of Personal Fall Arrest Systems

Anchorages

Body harness

Components (connectors like snaphooks or Dee-rings, connection points, lanyards, deceleration devices, lifelines, etc.)

- Limit the maximum arresting force on a worker to 1,800 pounds when used with a body harness. 29 CFR 1926.502(d)(16)(ii).

- Be rigged so that a worker can neither free fall more than 6 feet nor contact any lower level. 29 CFR 926.502(d)(16)(iii).

- Bring a worker to a complete stop and limit the maximum deceleration distance a worker travels to 3.5 feet. 29 CFR 1926.502(d)(16)(iv).

- Have sufficient strength to withstand twice the potential impact energy of a worker free falling a distance of 6 feet or the free fall distance permitted by the system, whichever is less. 29 CFR 1926.502(d)(16)(v).

- Be inspected prior to each use for wear, damage, and other deterioration. Defective components must be removed from service. 29 CFR 1926.502(d)(21).

***Note**: Limited use of body belts (safety belts) can still be used as part of a positioning device system or fall restraint system See more information under Positioning Device Systems and Fall Restraint Systems, below.

Personal Fall Arrest System Components

Snaphooks

- Snaphooks must be the locking type and designed and used to prevent disengagement from any component part of the personal fall arrest system. 29 CFR 1926.502(d)(5).
- Locking type snaphooks may also be used when designed for the following connections:
  - directly to webbing, rope, or wire rope;
  - to each other;
  - to a Dee-ring to which another snaphook or other connector is attached;
  - to a horizontal lifeline; or
  - to any object which is incompatibly shaped or dimensioned in relation to the snaphook, such that unintentional disengagement could occur by the connected object being able to depress the snaphook keeper and release itself. 29 CFR 1926.502(d)(6).

Horizontal Lifelines

- On suspended scaffolds or similar work platforms with horizontal lifelines that may become vertical lifelines, the devices used to connect to a horizontal lifeline must be capable of locking in both directions on the lifeline. 29 CFR 1926.502(d)(7).
- Horizontal lifelines must be designed, installed, and used under the supervision of a qualified person, as part of a complete personal fall arrest system that maintains a safety factor of at least two. 29 CFR 1926.502(d)(8).

Vertical Lifelines and Lanyards

■ Vertical lifelines and lanyards must have a minimum breaking strength of 5,000 pounds. 29 CFR 1926.502(d)(9).

■ Lifelines must be protected against being cut or abraded. 29 CFR 1926.502(d)(11).

Self-retracting Lifelines and Lanyards

■ Self-retracting lifelines and lanyards that automatically limit free fall distance to 2 feet or less must be capable of sustaining a minimum tensile load of 3,000 pounds applied to the device with the lifeline or lanyard in the fully extended position. 29 CFR 1926.502(d)(12).

■ Self-retracting lifelines and lanyards which do not limit free fall distance to 2 feet or less, ripstitch lanyards, and tearing and deforming lanyards must be capable of sustaining a minimum tensile load of 5,000 pounds applied to the device with the lifeline or lanyard in the fully extended position. 29 CFR 1926.502(d)(13).

Ropes and Straps

■ Ropes and straps (webbing) used in lanyards, lifelines, and strength components of body belts and body harnesses must be made of synthetic fibers. 29 CFR 1926.502(d)(14).

Anchorages

■ Anchorages used to attach personal fall arrest systems must be designed, installed, and used under the supervision of a qualified person, as part of a complete personal fall arrest system which maintains a safety factor of at least two. Alternatively, the anchorages must be independent of any anchorage being used to support or suspend platforms and must be capable of supporting at least 5,000 pounds per worker attached or be capable of supporting at least twice the expected impact load. 29 CFR 1926.502(d)(15).

Positioning Device Systems – 29 CFR 1926.502(e)

OSHA defines a positioning device system as a body belt or body harness system rigged to allow a worker to be supported on an elevated vertical surface, such as a wall, and work with both hands free while leaning.

- Body belt or body harness systems are to be set up so that a worker can free fall no farther than 2 feet. 29 CFR 1926.502(e)(1).
- Body belts or harnesses must be secured to an anchorage capable of supporting at least twice the potential impact load of a worker's fall or 3,000 pounds, whichever is greater. 29 CFR 1926.502(e)(2).

Positioning Device System Components

Snaphooks, Dee-rings, and Other Connectors

Requirements for components are similar or identical to provisions relating to Personal Fall Arrest System components found in 29 CFR 1926.502(d).

For strength and safe use requirements of snaphooks, Dee-rings, and other connectors when used with positioning device systems, see 29 CFR 1926.502(e)(3) through (10).

Fall Restraint Systems

While fall restraint systems are not mentioned in Subpart M, OSHA recognizes a fall restraint system as a means of prevention. The system, if properly used, tethers a worker in a manner that **will not allow a fall of any distance**. This system is comprised of a body belt or body harness, an anchorage, connectors, and other necessary equipment. Other components typically include

Photo: Skip Pennington

a lanyard, a lifeline, and other devices. For a restraint system to work, the anchorage must be strong enough to prevent the worker from moving past the point where the system is fully extended, including an appropriate safety factor.

*In a November 2, 1995 interpretation letter to Mr. Dennis Gilmore, OSHA suggested that, at a minimum, a fall restraint system must have the capacity to withstand at least 3,000 pounds or twice the maximum expected force that is needed to restrain the person from exposure to the fall hazard. In determining this force, consideration should be given to site-specific factors such as the force generated by a person (including his/her tools, equipment, and materials) walking, slipping, tripping, leaning, or sliding along the work surface.

*Letters of Interpretation

There are a number of Letters of Interpretation pertinent to fall protection that may affect your operation. OSHA Letters of Interpretation do not create new or additional requirements but rather explain these requirements and how they apply to particular circumstances. The letters constitute OSHA's interpretation of the requirements discussed. From time to time, letters are affected when the Agency updates a standard, a legal decision impacts a standard, or changes in technology affect the interpretation. To assure that you are using the correct information and guidance, please consult OSHA's website at www.osha.gov. If you have further questions, contact the Directorate of Construction at (202) 693-2020.

Additional Fall Protection Systems

Warning Line Systems – 29 CFR 1926.502(f)

OSHA defines a warning line system as a barrier erected on a roof to warn workers that they are approaching an unprotected roof side or edge, and to designate an area in which roofing work may take place without the use of guardrails, body harnesses, or safety net systems to protect workers in the area. Warning line systems consist of ropes, wires, or chains, plus supporting stanchions. If an employer chooses to use warning line systems, the following provisions apply:

- The warning line must be erected around all sides of roof work areas. 29 CFR 1926.502(f)(1).
- When mechanical equipment is not being used, the warning line must be erected at least 6 feet from the roof edge. 29 CFR 1926.502(f)(1)(i).
- When mechanical equipment is being used the warning line must be erected:
  - At least 6 feet from the roof edge parallel to the direction of mechanical equipment operation; and
  - At least 10 feet from the roof edge perpendicular to the direction of mechanical equipment operation. 29 CFR 1926.502(f)(1)(ii).
- The rope, wire, or chain must be flagged at not more than 6-foot intervals with high-visibility material. 29 CFR 1926.502(f)(2)(i).
- The rope, wire, or chain must be rigged and supported so that:
  - The lowest point (including sag) is at least 34 inches from the walking or working surface; and
  - Its highest point is no more than 39 inches from the walking or working surface. 29 CFR 1926.502(f)(2)(ii).
- Stanchions, after being rigged with warning lines, must be capable of resisting, without tipping over, a force of at least 16 pounds applied horizontally against the stanchion, 30 inches above the walking or working surface, perpendicular to the warning line and in the direction of the floor, roof, or platform edge. 29 CFR 1926.502(f)(2)(iii).
- The rope, wire, or chain must have a minimum tensile strength of 500 pounds. After being attached to the stanchions, it must support, without breaking, the loads applied to the stanchions as prescribed in 29 CFR 1926.502(f)(2)(iii) & 29 CFR 1926.502(f)(2)(iv).

■ The rope, wire, or chain must be attached to each stanchion in such a way that pulling on one section of the line between stanchions will not result in slack being taken up in the adjacent section before the stanchion tips over. 29 CFR 1926.502(f)(2)(v).

Controlled Access Zones – 29 CFR 1926.502(g)

■ A controlled access zone is a work area in which certain types of work may take place without using conventional fall protection systems. Worker access to these areas must be carefully controlled. For example, a controlled access zone would be designated where overhand bricklaying was occurring without the protection of guardrails. In this example, only masons and other workers actually engaged in the bricklaying would be allowed in the controlled access zone.

■ When used to control access to areas where leading edge and other operations are taking place, the controlled access zones must be defined by a control line or by any other means that restricts access. 29 CFR 1926.502(g)(1).

■ When control lines are used to define a controlled access zone, they must be erected at least 6 feet and no more than 25 feet from the unprotected or leading edge, except when precast concrete members are being erected. In the latter case, the control line is to be erected at least 6 feet and no more than 60 feet or half the length of the member being erected, whichever is less, from the leading edge. 29 CFR 1926.502(g)(1)(i) and (ii).

■ The control line must extend along the entire length of and approximately parallel to the unprotected side or leading edge and be connected on each side to a guardrail system or wall. 29 CFR 1926.502(g)(1)(iii) and (iv).

■ When controlled access zones are used to limit access to areas where overhand bricklaying and related work are taking place:

  ○ A control line must be erected to define the work zone and must be erected at least 10 feet and no more than 15 feet from the working edge. 29 CFR 1926.502(g)(2)(i).

The control lines must be erected approximately parallel to the working edge and must extend for a distance sufficient to enclose all workers performing overhand bricklaying and related work at the working edge. 29 CFR 1926.502(g)(2)(ii).

- Additional control lines must be erected at each end of the controlled access zone to enclose the work area. 29 CFR 1926.502(g)(2)(iii).
- Only workers engaged in overhand bricklaying or related work are permitted in the controlled access zone. 29 CFR 1926.502(g)(2)(iv).

- Control lines must consist of ropes, wires, tapes, or equivalent materials, and supporting stanchions. When used, each control line must:
  - Be flagged or otherwise clearly marked at not more than 6-foot intervals with high-visibility material. 29 CFR 1926.502(g)(3)(i).
  - Be rigged and supported in such a way that the lowest point (including sag) is not less than 39 inches from the walking or working surface; and the highest point is not more than 45 inches, or more than 50 inches when overhand bricklaying operations are being performed, from the walking or working surface. 29 CFR 1926.502(g)(3)(ii).
  - Have a breaking strength of at least 200 pounds. 29 CFR 1926.502(g)(3)(iii).

- On floors and roofs where guardrail systems are not in place prior to the beginning of overhand bricklaying operations, controlled access zones must be enlarged as necessary to enclose all points of access, material handling areas, and storage areas. 29 CFR 1926.502(g)(4).
- On floors and roofs where guardrail systems are in place but need to be removed to allow overhand bricklaying work or leading edge work to take place, only that portion of the guardrail necessary to accomplish that day's work is allowed to be removed. 29 CFR 1926.502(g)(5).

Safety Monitoring Systems – 29 CFR 1926.502(h)

A safety monitoring system is an alternative fall protection option for low-slope roofing work under 29 CFR 1926.501(b)(10). If employers elect to use a safety monitoring system, they must designate a competent person to monitor the safety of workers and to warn them when their work puts them close to a fall hazard.

The safety monitor must:

- Be competent in the recognition of fall hazards. 29 CFR 1926.502(h)(1)(i);
- Warn workers when it appears that they are unaware of fall hazards or when the workers are acting in an unsafe manner. 29 CFR 1926.502(h)(1)(ii);
- Be on the same walking or working surfaces as the workers and be able to see them. 29 CFR 1926.502(h)(1)(iii);
- Be close enough to the work operations to speak directly with workers. 29 CFR 1926.502(h)(1)(iv); and
- Have no other duties to distract them from their monitoring function. 29 CFR 1926.502(h)(1)(v).

Employers must ensure that:

- Mechanical equipment is not used or stored in areas where safety monitoring systems are being used to monitor workers engaged in roofing operations on low-slope roofs. 29 CFR 1926.502(h)(2);
- No worker, other than one engaged in roofing work on low-slope roofs or one covered by a fall protection plan, enters an area where a worker is being protected by a safety monitoring system. 29 CFR 1926.502(h)(3); and
- All workers in a controlled access zone have been instructed to promptly comply with fall hazard warnings issued by safety monitors. 29 CFR 1926.502(h)(4).

Safety monitoring systems must also be used as part of a fall protection plan under 29 CFR 1926.502(k), where no other alternative measure has been implemented. As explained in more detail below, the use of a fall protection plan is limited to residential construction work, precast concrete work, and leading edge work (see 29 CFR 1926.501(b)(2), (b)(12) and (b)(13)). The employer must first demonstrate that it is infeasible or creates a greater hazard to use conventional fall protection equipment.

Other Hazards that Require Fall Protection

A construction environment poses many hazards requiring protection. Below are some fall hazards that cannot be overlooked.

Hoist Areas – 29 CFR 1926.501(b)(3)

Each worker in a hoist area must be protected from falling 6 feet or more by guardrail systems or personal fall arrest systems. There may be times when the guardrail systems (or chain, gate, or guardrail) must be removed in whole or part to facilitate hoisting operations. For example, during the landing of materials, a worker may need to lean through the access opening or out over the edge of the access opening to receive or guide equipment and materials. At such times a personal fall arrest system must be used to protect the worker from falling through the unprotected opening.

Holes – 29 CFR 1926.501(b)(4)

- Each worker on walking or working surfaces must be protected from falling through holes (including skylights) that are more than 6 feet above lower levels, by personal fall arrest systems, covers, or guardrail systems erected around such holes. 29 CFR 1926.501(b)(4)(i).
- Each worker on a walking or working surface must be protected from tripping in or stepping into or through holes (including skylights) by covers. 29 CFR 1926.501(b)(4)(ii).

Covered floor hole marked, and with a guardrail surrounding it.

Ramps, Runways, and Other Walkways – 29 CFR 1926.501(b)(6)

Each worker on a ramp, runway, or other walkway must be protected by guardrail systems against falling 6 feet or more.

Excavations – 29 CFR 1926.501(b)(7)

- Each worker at the edge of an excavation 6 feet or more deep must be protected from falling by guardrail systems, fences, or barricades when the excavation cannot be readily seen because of plant growth or other visual barrier. 29 CFR 1926.501(b)(7)(i).
- Each worker at the edge of a well, pit, shaft, and similar excavation 6 feet or more deep must be protected from falling by guardrail systems, fences, or barricades, or covers. 29 CFR 1926.502(b)(7)(ii).

Dangerous Equipment – 29 CFR 1926.501(b)(8)

- When working 6 feet or more above dangerous equipment, each worker must be protected by guardrail systems, safety net systems, or personal fall arrest systems. 29 CFR 1926.502(b)(8)(ii).
- When working _less_ than 6 feet above dangerous equipment, each worker must be protected from falling into or onto the dangerous equipment by a guardrail system or equipment guards. 29 CFR 1926.502(b)(8)(i).

Wall Openings – 29 CFR 1926.501(b)(14)

Each worker working on, at, above, or near wall openings (including those with chutes attached), where the outside bottom edge of the wall opening is 6 feet or more above lower levels and the inside bottom edge of the wall opening is less than 39 inches above the walking or working surface, must be protected with a guardrail system, a safety net system, or a personal fall arrest system.

Protection from Falling Objects

Falling objects can also pose a hazard to workers. Falling object protection must comply with the following provisions:

Guardrails – 29 CFR 1926.502(j)(5)

When guardrail systems are used to prevent materials from falling from one level to another, any openings must be small enough to prevent passage of falling objects.

Overhand Bricklaying and Related Work – 29 CFR 1926.502(j)(6)

During overhand bricklaying and related work, no materials or equipment except masonry and mortar may be stored within 4 feet of working edges. Excess mortar, broken or scattered masonry units, and all other materials and debris must be kept clear of the working area by removal at regular intervals.

Roofing Work – 29 CFR 1926.502(j)(7)

During roofing work, materials and equipment must not be stored within 6 feet of a roof edge unless guardrail systems are erected at the edge. Any materials piled, grouped, or stacked near a roof edge must be stable and self-supporting.

Toeboards – 29 CFR 1926.502(j)(1) through (4)

When toeboards are used as protection from falling objects, they must be erected along the edges of the overhead walking or working surface for a distance sufficient to protect workers working below. 29 CFR 1926.502(j)(1). Other criteria include:

- Toeboards must be capable of withstanding, without failure, a force of at least 50 pounds applied in any downward or outward direction at any point along the toeboard. 29 CFR 1926.502(j)(2).
- Toeboards must be at least 3.5 inches tall from their top edge to the level of the walking or working surface, must have no

more than 0.25 inch clearance above the walking or working surface, and must be solid or have openings no larger than one inch in its greatest dimension. 29 CFR 1926.502(j)(3).

- Where tools, equipment, or materials are piled higher than the top edge of a toeboard, paneling or screening must be erected from the walking or working surface or toeboard to the top of a guardrail system's top rail or midrail, for a distance sufficient to protect workers below. 29 CFR 1926.502(j)(4).

Canopies – 29 CFR 1926.502(j)(8)

When used as protection from falling objects, canopies must be strong enough to prevent collapse and to prevent penetration by any objects that may fall onto them.

Fall Protection Plans

Presumption of Feasibility

As a general matter, OSHA presumes that using conventional fall protection (that is, guardrails, personal fall arrest systems, or safety nets) is feasible and will not create a greater hazard to use. However, as outlined below, there are a few circumstances when an employer can use a site-specifc fall protection plan instead of conventional fall protection.

When Can I Use a Fall Protection Plan?

It is possible that during leading edge work (29 CFR 1926.501(b)(2)), precast concrete erection (29 CFR 1926.501(b)(12)), or residential construction (29 CFR 1926.501(b)(13)), it may be infeasible or may create a greater hazard to use conventional fall protection for a specific task. In those circumstances, employers may implement a fall protection plan that complies with 29 CFR 1926.502(k).

> **IMPORTANT:** The employer has the burden of establishing that it is appropriate to implement a fall protection plan instead of implementing conventional fall protection systems.

Elements of a Fall Protection Plan – 29 CFR 1926.502(k)

- A fall protection plan must be prepared by a qualified person and developed specifically for the site where the work is being performed. 29 CFR 1926.502(k)(1).

- The fall protection plan must be maintained and kept up to date. 29 CFR 1926.502(k)(1).

- Any changes to the fall protection plan must be approved by a qualified person. 29 CFR 1926.502(k)(2).

- A copy of the fall protection plan with all approved changes must be maintained at the job site. 29 CFR 1926.502(k)(3).

- A competent person must supervise the implementation of the fall protection plan. 29 CFR 1926.502(k)(4).

- The plan must document the reasons why the use of conventional fall protection is infeasible or would create a greater hazard. 29 CFR 1926.502(k)(5).

- The plan must include a written discussion of other measures that will be taken to reduce or eliminate the fall hazard for workers who cannot be provided with protection using conventional fall protection systems. For example, the employer must discuss the extent to which scaffolds, ladders, or vehicle-mounted work platforms can be used to provide a safer working surface and thereby reduce the hazard of falling. 29 CFR 1926.502(k)(6).

- The plan must identify each location where conventional fall protection methods cannot be used. These locations must then be classified as controlled access zones, and the employer must comply with the criteria in 29 CFR 1926.502(g) and 29 CFR 1926.502(k)(7).

- Where no other alternative measure has been implemented, the employer must implement a safety monitoring system that complies with 29 CFR 1926.502(h) and 29 CFR 1926.502(k)(8).

- The plan must include a statement which provides the name or other method of identification for each worker who is authorized to work in controlled access zones. No other workers may enter controlled access zones. 29 CFR 1926.502(k)(9).

■ In the event that a worker falls, or some other related, serious incident occurs (for example, a near miss), the employer must investigate the circumstances to determine if the fall protection plan needs to be changed. For example, the plan may need to add new practices, procedures, or training. The employer must implement the needed changes to prevent similar types of falls or incidents. 29 CFR 1926.502(k)(10).

Fall Protection Training

Requirements – 29 CFR 1926.503

Employers must provide a fall protection training program to workers who might be exposed to fall hazards. Training must include how to recognize fall hazards and how to minimize them. 29 CFR 1926.503(a)(1).

The employer must assure that each worker has been trained as necessary, by a competent person who is qualified in the following areas:

■ The nature of fall hazards in the work area. 29 CFR 1926.503(a)(2)(i).

■ The correct procedures for erecting, maintaining, disassembling, and inspecting the fall protection systems to be used. 29 CFR 1926.503(a)(2)(ii).

■ The use and operation of controlled access zones; guardrail, personal fall arrest, safety net, warning line, and safety monitoring systems; and other protection to be used. 29 CFR 1926.503(a)(2)(iii).

■ The role of each worker in the safety monitoring system when the system is used. 29 CFR 1926.503(a)(2)(iv).

■ The limitations on the use of mechanical equipment during the performance of roofing work on low-slope roofs. 29 CFR 1926.503(a)(2)(v).

■ The correct procedures for equipment and materials handling and storage and the erection of overhead protection. 29 CFR 1926.503(a)(2)(vi).

- The role of workers in fall protection plans. 29 CFR 1926.503(a)(2)(vii).
- OSHA's fall protection requirements, published as Subpart M. 29 CFR 1926.503(a)(2)(viii).

Verification of Training

Employers must verify worker training by preparing a <u>written</u> certification record. The record must contain the name or other identity of the worker trained, the dates of the training, and the signature of either the person who conducted the training or the employer. 29 CFR 1926.503(b)(1).

When an employer has reason to believe that an affected worker does not recognize existing fall hazards at some point after the initial training, the employer is required to provide retraining for that worker. For example, workers must be retrained when:

- Changes in the workplace render previous training obsolete. 29 CFR 1926.503(c)(1).
- Fall protection equipment or systems have changed. 29 CFR 1926.503(c)(2).
- Inadequacies in workers' knowledge or use of fall protection systems or equipment indicate that they have not adequately understood or retained previous training. 29 CFR 1926.503(c)(3).

Fall Protection Requirements in Other OSHA Construction Standards

Other OSHA construction standards also contain fall protection requirements. Employers covered by another standard may have to comply with the requirements in the other standard and those in Subpart M, unless one of the exceptions listed in 29 CFR 1926.500(a)(2), 1926.500(a)(3), or 1926.500(a)(4) applies. OSHA included these exceptions because other means of providing fall protection (for example, when using ladders and scaffolds) may eliminate the need for providing fall protection under Subpart M. For example, employers with

workers engaged in the construction of electric transmission
or distribution lines or equipment should refer to Subpart V –
Power Transmission and Distribution for specific fall protection
requirements. However, employers should be aware that their
workers are still covered by Subpart M when the workers are
engaged in activities not covered by Subpart V or another of
the exceptions under 29 CFR 1926.500.

The following subparts of OSHA's Construction standards
address fall protection requirements and performance criteria
outside of Subpart M:

- Personal Protective and Life Saving Equipment – Subpart E
  (applies to belts, lanyards, lifelines, nets for work on tanks,
  communication and broadcast towers);
- Scaffolds – Subpart L;
- Steel Erection – Subpart R;
- Underground Construction, Caissons, Cofferdams, and
  Compressed Air – Subpart S (applies to certain types of
  equipment in tunneling operations);
- Power Transmission and Distribution – Subpart V;
- Stairways and Ladders – Subpart X; and
- Cranes and Derricks in Construction – Subpart CC.

Subpart M – Fall Protection: Non-mandatory Appendices

Appendix A to Subpart M – Determining Roof Widths – Non-mandatory Guidelines for Complying with 1926.501(b)(10)

This appendix serves as a guideline to help employers comply
with the requirements of 1926.501(b)(10). Section 1926.501(b)(10)
allows the use of a safety monitoring system alone as a means
of providing fall protection during the performance of roofing
operations on low-sloped roofs 50 feet (15.25 m) or less in width.

Appendix B to Subpart M – Guardrail Systems – Non-mandatory Guidelines for Complying with 1926.502(b)

This appendix serves as a guideline to assist employers in
designing and building guardrail systems in compliance with
1926.502(b)(3), (4), and (5).

Appendix C to Subpart M – Personal Fall Arrest Systems – Non-mandatory Guidelines for Complying with 1926.502(d) This appendix serves as a non-mandatory guideline to help employers comply with the requirements in 1926.502(d). Section I, paragraphs (b), (c), (d) and (e) describe methods for testing personal fall arrest systems and positioning device systems.

Appendix D to Subpart M – Positioning Device Systems – Non-mandatory Guidelines for Complying with 1926.502(e) This appendix serves as a non-mandatory guideline to help employers comply with the requirements for positioning device systems in 1926.502(e). The procedures listed here, along with those in Appendix C (above), describe methods for testing positioning device systems to comply with 1926.502(e) (3) and (4) of Subpart M.

Appendix E to Subpart M – Sample Fall Protection Plan – Non-mandatory Guidelines for Complying with 1926.502(k) This appendix provides sample fall protection plans for employers engaged in leading edge work, precast concrete construction, or residential construction work. The employer must be able to demonstrate that it is infeasible or creates a greater hazard to use conventional fall protection systems.

Definitions

The definitions in this section that affect Subpart M are found in 29 CFR 1926.32; 29 CFR 1926.500; and STD 03-11-002 Compliance Guidance for Residential Construction.

Anchorage – a secure point of attachment for lifelines, lanyards, or deceleration devices.

Body belt (safety belt) – a strap with means both for securing it about the waist and for attaching it to a lanyard, lifeline, or deceleration device. *Note: Since January 1, 1998, OSHA has prohibited the use of a body belt as part of a **personal fall arrest system**. Exception: When used correctly, body belts are recognized by OSHA as an acceptable fall protection component when used as a part of either a restraining device which prevents a fall or a positioning device which limits a free fall to 2 feet.*

Body harness – straps which may be secured about the worker in a manner that will distribute the fall arrest forces over at least the thighs, pelvis, waist, chest, and shoulders, with means for attaching it to other components of a personal fall arrest system.

Buckle – any device for holding the body belt or body harness closed around the worker's body.

Competent person – one who is capable of identifying existing and predictable hazards in the surroundings or working conditions which are unsanitary, hazardous, or dangerous to workers, and who has authorization to take prompt corrective measures to eliminate them.

Connector – a device which is used to couple (connect) parts of the personal fall arrest system and positioning device systems together. It may be an independent component of the system, such as a carabiner, or it may be an integral component of part of the system (such as a buckle or Dee-ring sewn into a body belt or body harness, or a snaphook spliced or sewn to a lanyard or self-retracting lanyard).

Controlled access zone (CAZ) – an area in which certain work (for example, overhand bricklaying) may take place without the use of guardrail systems, personal fall arrest systems, or safety net systems; and where access to the zone is controlled.

Dangerous equipment – equipment (such as pickling or galvanizing tanks, degreasing units, machinery, electrical equipment, and other units) which, as a result of form or function, may be hazardous to workers who fall onto or into such equipment.

Deceleration device – any mechanism (such as a rope grab, rip-stitch lanyard, specially-woven lanyard, tearing or deforming lanyards, automatic self-retracting lifelines/lanyards, etc.) which serves to dissipate a substantial amount of energy during a fall arrest, or otherwise limit the energy imposed on a worker during fall arrest.

Deceleration distance – the additional vertical distance a falling employee travels, excluding lifeline elongation and free fall distance, before stopping, from the point at which the deceleration device begins to operate. It is measured as the distance between the location of a worker's body belt or body harness attachment point at the moment of activation (at the onset of fall arrest forces) of the deceleration device during a fall, and the location of that attachment point after the worker comes to a full stop.

Equivalent – alternative designs, materials, or methods to protect against a hazard, which the employer can demonstrate will provide an equal or greater degree of safety for workers than the methods, materials, or designs specified in the standard.

Failure – load refusal, breakage, or separation of component parts. Load refusal is the point where the ultimate strength is exceeded.

Free fall – the act of falling before a personal fall arrest system begins to apply force to arrest the fall.

Free fall distance – the vertical displacement of the fall arrest attachment point on the worker's body belt or body harness between onset of the fall and just before the system begins to apply force to arrest the fall. This distance excludes deceleration distance and lifeline/lanyard elongation, but includes any deceleration device slide distance or self-retracting lifeline/lanyard extension before they operate and fall arrest forces occur.

Guardrail system – a barrier erected to prevent workers from falling to lower levels.

Hole – a gap or void 2 inches or more in its least dimension, in a floor, roof, or other walking or working surface.

Infeasible – impossible to perform the construction work using a conventional fall protection system (that is, guardrail system, safety net system, or personal fall arrest system); or technologically impossible to use any one of these systems to provide fall protection.

Lanyard – a flexible line of rope, wire rope, or strap which generally has a connector at each end for connecting the body belt or body harness to a deceleration device, lifeline, or anchorage.

Leading edge – the edge of a floor, roof, or formwork for a floor or other walking or working surface (such as the deck) which changes location as additional floor, roof, decking, or formwork sections are placed, formed, or constructed. A leading edge is considered to be an "unprotected side and edge" during periods when it is not actively and continuously under construction.

Lifeline – a component consisting of a flexible line for connection to an anchorage at one end to hang vertically (vertical lifeline), or for connection to anchorages at both ends to stretch horizontally (horizontal lifeline), and which serves as a means for connecting other components of a personal fall arrest system to the anchorage.

Low-slope roof – a roof having a slope less than or equal to 4 in 12 (vertical to horizontal).

Lower levels – those areas or surfaces to which a worker can fall. Such areas or surfaces include, but are not limited to, ground levels, floors, platforms, ramps, runways, excavations, pits, tanks, material, water, equipment, structures, or portions thereof.

Mechanical equipment – all motor- or human-propelled wheeled equipment used for roofing work, except wheelbarrows and mop carts.

Opening – a gap or void 30 inches or more high and 18 inches or more wide, in a wall or partition, through which workers can fall to a lower level.

Overhand bricklaying and related work – the process of laying bricks and masonry units such that the surface of the wall to be jointed is on the opposite side of the wall from the mason, requiring the mason to lean over the wall to complete the work. Related work includes mason tending and electrical installation incorporated into the brick wall during the overhand bricklaying process.

Personal fall arrest system – a system used to arrest a worker in a fall from a working level. It consists of an anchorage, connectors, and a body harness. It may include a lanyard, deceleration device, lifeline, or suitable combinations of these. *Note: Since January 1, 1998, the use of a body belt for fall arrest has been prohibited.*

Positioning device system – a body belt or body harness system rigged to allow a worker to be supported on an elevated vertical surface, such as a wall, and work with both hands free while leaning.

Qualified – one who, by possession of a recognized degree, certificate, or professional standing, or who by extensive knowledge, training, and experience, has successfully demonstrated his ability to solve or resolve problems relating to the subject matter, the work, or the project.

Rope grab – a deceleration device which travels on a lifeline and automatically, by friction, engages the lifeline and locks so as to arrest the fall of a worker. A rope grab usually employs the principles of inertial locking, cam/level locking, or both.

Roof – the exterior surface on the top of a building. This does not include floors or formwork which, because a building has not been completed, temporarily become the top surface of a building.

Roofing work – the hoisting, storage, application, and removal of roofing materials and equipment, including related insulation, sheet metal, and vapor barrier work, but not including the construction of the roof deck.

Safety-monitoring system – a safety system in which a competent person is responsible for recognizing and warning workers of fall hazards.

Self-retracting lifeline/lanyard – a deceleration device containing a drum-wound line which can be slowly extracted from, or retracted onto, the drum under slight tension during normal worker movement, and which, after onset of a fall, automatically locks the drum and arrests the fall.

Snaphook – a connector comprised of a hook-shaped member with a normally closed keeper, or similar arrangement, which may be opened to permit the hook to receive an object and, when released, automatically closes to retain the object. Snaphooks are generally one of two types:

(1) The *locking* type with a self-closing, self-locking keeper which remains closed and locked until unlocked and pressed open for connection or disconnection; or

(2) The *non-locking* type with a self-closing keeper which remains closed until pressed open for connection or disconnection. As of January 1, 1998, the use of a non-locking snaphook as part of personal fall arrest systems and positioning device systems is prohibited.

Steep roof – a roof having a slope greater than 4 in 12 (vertical to horizontal).

Toeboard – a low protective barrier that will prevent the fall of materials and equipment to lower levels and provide workers protection from falls.

Unprotected sides and edges – any side or edge (except at entrances to points of access) of a walking or working surface (for example, floor, roof, ramp, or runway) where there is no wall or guardrail system at least 39 inches high.

Walking/working (walking or working) surface – any surface (whether horizontal or vertical) on which a worker walks or works, including but not limited to floors, roofs, ramps, bridges, runways, formwork and concrete reinforcing steel; but not including ladders, vehicles, or trailers, on which workers must be located in order to perform their job duties.

Warning line system – a barrier erected on a roof to warn workers that they are approaching an unprotected roof side or edge, and which designates an area in which roofing work may take place without the use of guardrail, body harness, or safety net systems to protect workers in the area.

Work area – that portion of a walking or working surface where job duties are being performed.

Workers' Rights

Workers have the right to:

- Working conditions that do not pose a risk of serious harm.
- Receive information and training (in a language and vocabulary the worker understands) about workplace hazards, methods to prevent them, and the OSHA standards that apply to their workplace.
- Review records of work-related injuries and illnesses.
- File a complaint asking OSHA to inspect their workplace if they believe there is a serious hazard or that their employer is not following OSHA's rules. OSHA will keep all identities confidential.
- Exercise their rights under the law without retaliation, including reporting an injury or raising health and safety concerns with their employer or OSHA. If a worker has been retaliated against for using their rights, they must file a complaint with OSHA as soon as possible, but no later than 30 days.

For more information, see OSHA's Workers page.

OSHA Assistance, Services and Programs

OSHA has a great deal of information to assist employers in complying with their responsibilities under OSHA law. Several OSHA programs and services can help employers identify and correct job hazards, as well as improve their injury and illness prevention program.

Establishing an Injury and Illness Prevention Program

The key to a safe and healthful work environment is a comprehensive injury and illness prevention program.

Injury and illness prevention programs are systems that can substantially reduce the number and severity of workplace injuries and illnesses, while reducing costs to employers. Thousands of employers across the United States already manage safety using injury and illness prevention programs, and OSHA believes that all employers can and should do the same. Thirty-four states have requirements or voluntary guidelines for workplace injury and illness prevention programs. Most successful injury and illness prevention programs are based on a common set of key elements. These include management leadership, worker participation, hazard identification, hazard prevention and control, education and training, and program evaluation and improvement. Visit OSHA's Injury and Illness Prevention Programs web page at www.osha.gov/dsg/topics/safetyhealth for more information.

Compliance Assistance Specialists

OSHA has compliance assistance specialists throughout the nation located in most OSHA offices. Compliance assistance specialists can provide information to employers and workers about OSHA standards, short educational programs on specific hazards or OSHA rights and responsibilities, and information on additional compliance assistance resources. For more details, visit www.osha.gov/dcsp/compliance_assistance/cas.html or call 1-800-321-OSHA (6742) to contact your local OSHA office.

Free On-site Safety and Health Consultation Services for Small Business

OSHA's On-site Consultation Program offers free and confidential advice to small and medium-sized businesses in all states across the country, with priority given to high-hazard worksites. Each year, responding to requests from small employers looking to create or improve their safety and health management programs, OSHA's On-site Consultation Program conducts over 29,000 visits to small business worksites covering over 1.5 million workers across the nation.

On-site consultation services are separate from enforcement and do not result in penalties or citations. Consultants from state agencies or universities work with employers to identify workplace hazards, provide advice on compliance with OSHA standards, and assist in establishing safety and health management programs.

For more information, to find the local On-site Consultation office in your state, or to request a brochure on Consultation Services, visit www.osha.gov/consultation, or call 1-800-321-OSHA (6742).

Under the consultation program, certain exemplary employers may request participation in OSHA's **Safety and Health Achievement Recognition Program (SHARP)**. Eligibility for participation includes, but is not limited to, receiving a full-service, comprehensive consultation visit, correcting all identified hazards and developing an effective safety and health management program. Worksites that receive SHARP recognition are exempt from programmed inspections during the period that the SHARP certification is valid.

Occupational Safety and Health Training Courses

The OSHA Training Institute partners with 27 OSHA Training Institute Education Centers at 42 locations throughout the United States to deliver courses on OSHA standards and occupational safety and health topics to thousands of students a year. For more information on training courses, visit www.osha.gov/otiec.

OSHA Educational Materials

OSHA has many types of educational materials in English, Spanish, Vietnamese and other languages available in print or online. These include:

- Brochures/booklets;
- Fact Sheets;
- Guidance documents that provide detailed examinations of specific safety and health issues;
- Online Safety and Health Topics pages;
- Posters;
- Small, laminated QuickCards™ that provide brief safety and health information; and
- *QuickTakes*, OSHA's free, twice-monthly online newsletter with the latest news about OSHA initiatives and products to assist employers and workers in finding and preventing

workplace hazards. To sign up for *QuickTakes* visit www.osha.gov/quicktakes.

To view materials available online or for a listing of free publications, visit www.osha.gov/publications. You can also call 1-800-321-OSHA (6742) to order publications.

OSHA's web site also has information on job hazards and injury and illness prevention for employers and workers. To learn more about OSHA's safety and health resources online, visit www.osha.gov. Use the A-Z index to help find information and assistance.

NIOSH Health Hazard Evaluation Program

Getting Help with Health Hazards

The National Institute for Occupational Safety and Health (NIOSH) is a federal agency that conducts scientific and medical research on workers' safety and health. At no cost to employers or workers, NIOSH can help identify health hazards and recommend ways to reduce or eliminate those hazards in the workplace through its Health Hazard Evaluation (HHE) Program.

Workers, union representatives and employers can request a NIOSH HHE. An HHE is often requested when there is a higher than expected rate of a disease or injury in a group of workers. These situations may be the result of an unknown cause, a new hazard, or a mixture of sources. To request a NIOSH Health Hazard Evaluation go to www.cdc.gov/niosh/hhe/request.html. To find out more about the Health Hazard Evaluation Program:

- Call (513) 841-4382, or to talk to a staff member in Spanish, call (513) 841-4439; or
- Send an email to HHERequestHelp@cdc.gov.

OSHA Regional Offices

Region I
Boston Regional Office

(CT*, ME, MA, NH, RI, VT*)
JFK Federal Building, Room E340
Boston, MA 02203
(617) 565-9860 (617) 565-9827 Fax

Region II
New York Regional Office
(NJ*, NY*, PR*, VI*)
201 Varick Street, Room 670
New York, NY 10014
(212) 337-2378 (212) 337-2371 Fax

Region III
Philadelphia Regional Office
(DE, DC, MD*, PA, VA*, WV)
The Curtis Center
170 S. Independence Mall West
Suite 740 West
Philadelphia, PA 19106-3309
(215) 861-4900 (215) 861-4904 Fax

Region IV
Atlanta Regional Office
(AL, FL, GA, KY*, MS, NC*, SC*, TN*)
61 Forsyth Street, SW, Room 6T50
Atlanta, GA 30303
(678) 237-0400 (678) 237-0447 Fax

Region V
Chicago Regional Office
(IL*, IN*, MI*, MN*, OH, WI)
230 South Dearborn Street
Room 3244
Chicago, IL 60604
(312) 353-2220 (312) 353-7774 Fax

Region VI
Dallas Regional Office
(AR, LA, NM*, OK, TX)

525 Griffin Street, Room 602
Dallas, TX 75202
(972) 850-4145 (972) 850-4149 Fax
(972) 850-4150 FSO Fax

Region VII
Kansas City Regional Office
(IA*, KS, MO, NE)
Two Pershing Square Building
2300 Main Street, Suite 1010
Kansas City, MO 64108-2416
(816) 283-8745 (816) 283-0547 Fax

Region VIII
Denver Regional Office
(CO, MT, ND, SD, UT*, WY*)
Cesar Chavez Memorial Building
1244 Speer Boulevard, Suite 551
Denver, CO 80204
(720) 264-6550 (720) 264-6585 Fax

Region IX
San Francisco Regional Office
(AZ*, CA*, HI*, NV*, and American Samoa,
Guam and the Northern Mariana Islands)
90 7th Street, Suite 18100
San Francisco, CA 94103
(415) 625-2547 (415) 625-2534 Fax

Region X
Seattle Regional Office
(AK*, ID, OR*, WA*)
300 Fifth Avenue, Suite 1280
Seattle, WA 98104
(206) 757-6700 (206) 757-6705 Fax

* These states and territories operate their own OSHA-
approved job safety and health plans and cover state and local
government employees as well as private sector employees.
The Connecticut, Illinois, New Jersey, New York and Virgin

Islands programs cover public employees only. (Private sector workers in these states are covered by Federal OSHA). States with approved programs must have standards that are identical to, or at least as effective as, the Federal OSHA standards.

Note: To get contact information for OSHA area offices, OSHA-approved state plans and OSHA consultation projects, please visit us online at www.osha.gov or call us at 1-800-321-OSHA (6742).